Audel™
Welding Pocket Reference

T0131147

Audel™
Welding Pocket Reference

James E. Brumbaugh
Rex Miller

BICENTENNIAL
1807
WILEY
2007
BICENTENNIAL

Publisher: Joe Wikert
Production Editor: Felicia Robinson
Text Design & Composition: Techbooks

Library of Congress Cataloging-in-Publication Data.

ISBN: 10-0764588095

ISBN: 13-9780764588099

Printed in the United States of America

SKY10063990_010524

Table of Contents

Acknowledgments   *vii*
Preface   *ix*

1. Welding and Cutting Safety   1
2. Oxyacetylene Welding   13
3. Oxyacetylene Cutting   47
4. Shielded Metal Arc Welding   81
5. Arc Welding and Cutting   117
6. Brazing and Braze Welding   147
7. Soldering   165
8. Identifying Metals   199
9. Cast Iron   217
10. Wrought Iron   233
11. Carbon Steels   237
12. Alloy Steels   243
13. Reactive and Refractory Metals   255
14. Galvanized Metals   265
15. Soft Metals Welding   269
16. Magnesium and Magnesium Alloys   289
17. Lead, Tin and Zinc   301
18. Hard Facing and Tool and Die Steels   305
19. Types of Joints   319
20. Welding Positions   321
21. Types of Welds   327
22. Weld Terminology   333

23. Welding Symbols   341

24. Tips for Producing Good Welds   387

Appendix A   SMAW Carbon Steel Electrodes   389

Appendix B   SMAW Stainless Steel Electrodes   401

Appendix C   SMAW Aluminum Electrodes   411

Appendix D   Conversion Tables   415

Acknowledgments

No book of this nature can be written without the aid of many people. It takes a great number of individuals to put together the information available about any particular technical field into a book. The field of welding is no exception. Many firms have also contributed information, illustrations, and analysis of the book.

The authors would like to thank every person involved for his or her contributions. The following organizations are among those that supplied technical information and illustrations.

Airco Welding Products
American Welding Society
Aufhauser Corporation
James F. Lincoln Arc Welding Foundation
Rego Company
Stoody, a Thermodyne Company
Victor Equipment Company
Welder's Guide

Preface

Audel's *Welding Pocket Reference* is written for everyone who wants or needs to know about welding. Whether you are trying to construct a building, repair a railroad problem, or solder a tin can for a project, the rewards from a job well done are many.

This book can be used by students in vocational courses, technical colleges, apprenticeship programs, and welding classes in industrial technology programs. The home do-it-yourselfer will find answers to questions that pop up in the course of getting a job done, whether over a weekend or a longer period of time. The professional welder will find this to be a handy reference while working on a project in the field, and everyday welding problems can be handled easily with the aid of this book.

The purpose of this book is to aid you in your everyday tasks and keep you updated with the latest facts, figures, and devices in this important trade.

Many illustrations are included that show a variety of parts and techniques found in present-day practice in the field. Obviously, not all related problems can be presented here, and a worker needs a great deal of ingenuity on the job. For standard procedure, however, the *Welding Pocket Reference* does give a guide to which electrode numbers to use for particular metals, what welding tips to use, and other important facts and figures. This handbook should be kept in your pocket or toolbox for easy access.

Audel™
Welding Pocket Reference

I. WELDING AND CUTTING SAFETY

The *American Welding Society* and the *American Standards Association*, as well as various governmental and private organizations, have compiled safety standards for the operation of gas and arc welding and cutting equipment. You are encouraged to read these publications and become familiar with their recommendations. The various manufacturers of welding equipment and supplies also publish safety information.

OXYACETYLENE WELDING AND CUTTING SAFETY

The safety recommendations for oxyacetylene welding and cutting covered in this chapter also apply to the other oxyfuel processes, such as oxyhydrogen welding and cutting, oxy-natural gas cutting, and air-acetylene welding.

General Safety Recommendations for Oxyacetylene Welding and Cutting

- Never weld in the vicinity of flammable or combustible materials.
- Never weld on containers that have held combustible or flammable materials without first exercising the proper precautions.
- Never weld in confined spaces without adequate ventilation or individual respiratory equipment.
- Never pick up hot objects.
- Never do any chipping or grinding without suitable goggles.

General Recommendations for Safe Handling of Cylinders and Regulators

- Make certain that the connections between the regulators, adapters, and cylinder valves are gas-tight.

Escaping acetylene can generally be detected by the odor. Test with soapy water—never with an open flame.

- Never move individual cylinders unless the valve protection cap, where provided, is in place, hand-tight.
- Never drop or roughly handle cylinders in any way.
- Make certain that cylinders are well fastened in their station so that they will not fall.
- Never use a hammer or wrench to open any valve on a cylinder.
- Never force connections that do not fit.
- Never tamper with cylinder safety devices.
- Never use oil to clean a regulator gauge. As a combustible substance, oil has a very low flash point.

Protecting Oxygen and Acetylene Hoses

- Always protect hoses from being stepped on or run over. Avoid tangles and kinks. Never leave hoses where they can be tripped over.
- Protect hoses and cylinders from flying sparks, hot slag, hot objects, and open flame.
- Never allow a hose to come into contact with oil or grease; these substances deteriorate the rubber and constitute a hazard with oxygen.

Oxygen Gas and Cylinder Safety Recommendations

- Always refer to oxygen by its full name—oxygen—and not by the word air. This will avoid the possibility of confusing oxygen with compressed air.
- Never use oxygen near flammable materials, especially grease, oil, or any substance likely to cause or accelerate fire. Oxygen itself is not flammable, but it does *strongly* support combustion.
- Do not store oxygen and acetylene cylinders together. They should be separately grouped.

- Never permit oil or grease to come in contact with oxygen cylinders, valves, regulators, hoses, or fittings. Do not handle oxygen cylinders with oily hands or oily gloves.

- Never use oxygen pressure-reducing regulators, hoses, or other pieces of apparatus with any other gases.

- Always open the oxygen cylinder valve slowly.

- Never attempt to store any other gases in an oxygen cylinder.

- Oxygen must never be used for ventilation or as a substitute for compressed air.

- Never use oxygen from cylinders without first connecting a suitable pressure-reducing regulator to the cylinder valve.

- Never tamper with nor attempt to repair oxygen cylinder valves, unless qualified to do so.

Acetylene Fuel Gas and Cylinder Safety Recommendations

- Call acetylene by its full name—acetylene—and not by the word gas.

- Acetylene cylinders should be used and stored valve end up.

- Never use acetylene from cylinders without a suitable pressure valve.

- Turn the acetylene cylinder so that the valve outlet will point away from the oxygen cylinder.

- When opening an acetylene cylinder valve, do not turn the key or spindle more than one and one-half turns.

- The acetylene cylinder key for opening the valve must be kept on the valve stem while the cylinder is in use, so that the acetylene cylinder may be quickly turned off in an emergency.

- Never use acetylene pressure-reducing regulators, hoses, or other pieces of apparatus with any other gases.
- Never attempt to transfer acetylene from one cylinder to another, refill an acetylene cylinder, or store any other gases in an acetylene cylinder.
- Should a leak occur in an acetylene cylinder, take the cylinder out into the open air, keeping it well away from fires or open lights. Notify the manufacturer immediately if any leaks occur.
- Never use acetylene at pressures in excess of 15 psi. The use of higher pressures is prohibited by all insurance authorities, and by law in many localities.

General Welding and Cutting Safety Tips

- Never use matches for lighting torches (hand burns may result). Use spark lighters, stationary pilot flames, or some other suitable source of ignition.
- Do not light torches from hot work while in a pocket or small confined space.
- Never attempt to relight a torch that has blown out without first closing both valves and relighting in the proper manner.
- Never hang a torch and its hose on regulators or cylinder valves.
- Never cut material in a position that will permit sparks, hot metal, or the severed section to fall on the cylinder, hose, or your legs or feet.
- When welding or cutting is to be stopped temporarily, release the pressure-adjusting screws of the regulators by turning them to the left.
- When the welding or cutting is to be stopped for a long time (during lunch hour or overnight) or taken down, close the cylinder valves and then release all gas

pressures from the regulators and hose by opening the torch valves momentarily. Close the torch valves and release the pressure-adjusting screws. If the equipment is to be taken down, make certain that all gas pressures are released from the regulators and hoses and that the pressure-adjusting screws are turned to the left until free.

ARC WELDING AND CUTTING SAFETY

Important safety recommendations for the arc welding and cutting processes include the following:

- Keep the work area and floor clean and clear of electrode stubs, scraps of metal, and carelessly placed tools.
- Always work in a well-ventilated area. This is the best protection against toxic fumes and dust. If the ventilation is poor, adequate respiratory equipment is necessary.
- Make sure cable connections are tight and that cables do not become hot.
- Never look at an electric arc with the naked eye. An electric arc gives off harmful radiation. Goggles with suitable lenses and protective clothing are recommended as protection against these rays, as well as against flying sparks, splattering metal, and hot metal.
- Electric shock can be avoided by proper handling of the arc welding equipment. The arc welding machine must be properly grounded at all times. The work area should be dry. All insulation (wiring, electrodes, and so forth) should be checked, and replaced if damaged.
- Never weld while wearing wet gloves or wet shoes.
- Never use electrode holders with defective jaws.
- Never leave the electrode holder on the table or in contact with a grounded metallic surface. Place it on the support provided for that purpose.

- Never weld on closed containers or on containers that have held combustible materials.
- Never allow an arc welding machine to rest on a dirt floor.
- Operate arc welding machines and equipment only in clean, dry locations.
- Insofar as possible, protect arc welding machines in the field from weather conditions.
- Always install an arc welding machine in compliance with the requirements of the *National Electrical Code* and local ordinances, and make certain it is properly grounded.
- Use the proper terminals on the arc welding machine for the power line voltage connection.
- Never work on the wiring of an arc welding machine unless qualified to do so.

WELDING AND CUTTING SAFETY EQUIPMENT

Eye Protection

The welder and others associated with welding operations should be provided with glasses designed to provide maximum protection while affording adequate vision for proper welding technique. Until recently it was customary to assume that if the visible rays were cut down to a comfortable intensity, the ultraviolet and infrared rays were also reduced proportionately. The usual practice was to use alternate layers of red and blue glass to reduce the light intensity to a value consistent with the work being done. Such a procedure may or may not offer complete protection, because some glasses strongly absorb the visible rays while transmitting the harmful infrared or ultraviolet rays quite freely.

Scientific methods for testing protective lenses have been developed. These tests have resulted in the establishment of standards for lenses of various grades and have led to the development of glass formulas that have materially increased the protective qualities of lenses. Complete and positive protection for the eyes is now available. Suggested shades for various types of welding and joining operations are given in Table 1-1.

Table 1-1 Recommended Lens Shades for Various Welding and Cutting Operations

Operation	Shade Number
Torch Brazing	3 or 4
Oxyacetylene Cutting	
>1″ thick metal	3 or 4
1 to 6″ thick metal	4 or 5
6″ thick metal and thicker	5 or 6
Oxyfuel Gas Welding	
>1/8″	4 or 5
1/8 to 1/2″	5 or 6
1/2″ and up	6 or 8
Oxyacetylene Flame Spraying	5
Shielded Metal-Arc Welding (SMAW)	
up to 5/32″ electrodes	10
3/16- to 1/4″ electrodes	12
5/16″ electrodes and up	14
Gas Metal-Arc Welding (GMAW) – MIG Welding	
>60 amps	8 to 10
60–160 amp range	11
160–250 amp range	12
250 amps and above	14
Flux Cored-Arc welding (FCAW)	
>60 amps	8 to 10
60–160 amp range	11
160–250 amp range	12
250 amps and above	14

(*continued*)

Table 1-1 (*continued*)

Operation	Shade Number
Gas Tungsten-Arc Welding (GTAW) – TIG Welding	
>20 amps	5 to 9
20–100 amp range	10 to 11
100–400 amp range	12
400 amps and above	14
Plasma-Arc Welding	
>20 amps	8
20–100 amp range	10
100–400 amp range	12
400 amps and above	14
Soldering	2

Eye Protection Tips:

- Seeing white spots in your vision after you've stopped welding and removed your goggles is an indication that you need darker lenses.
- Failure to distinguish between a neutral and a carburizing oxyacetylene flame while wearing your goggles is an indication that the colored lenses you are wearing are too dark.
- If you cannot see the weld puddle while wearing your goggles, you need a lighter-colored lens.

NOTE

Federal specifications for welding lenses specify not only the percentage of rays transmitted by the various shade numbers, but also the thickness of the glass and its optical properties.

Proper lens selection is important. There is a vast difference in welding lenses, both as to their value and their effect on welding production. It is impossible to distinguish one lens from another by casual inspection. Scientific tests are required to determine their qualities. It is vitally important when selecting a welding lens to take into account the reputation of the manufacturer and their experience in the welding field.

NOTE

Install screens or barriers to protect nearby workers from flash and glare created by the welding or cutting process.

Nonspatter cover glass is a chemically treated glass that protects the lens from spatter yet allows maximum visibility. Spatter does not adhere to this special type of glass, prolonging its life to five or ten times that of ordinary glass and maintaining clear visibility. A soft cloth should be used for cleaning.

Goggles are available in a great number of different designs and types. Either glass or plastic lenses may be purchased, and the lenses themselves can be either clear or tinted. Most goggles have either round or rectangular lenses.

Face shields are also useful for protecting the face of the operator against flying sparks and other dangerous matter. They find widespread use in arc welding and cutting. Face shields are available in clear plastic or in different shades of green and can be purchased in several thicknesses. (ANSI requires a minimum thickness of 0.041 inch or more.) Note: Face shields provide limited impact and splash protection.

Helmets, safety caps, and other headgear have been designed for operators of welding and cutting equipment to protect the head against serious blows. These are available in many different designs and a great number of sizes. The helmets are made from either molded fiberglass or metal plate.

Respirators

It is absolutely necessary that proper ventilation be provided for each welding and cutting operation. The fumes produced during welding and cutting can be injurious to the welder's health (see Table 1-2). Some fumes, such as those produced

Table 1-2 Some Common Sources of Potentially Hazardous Fumes

Type	Comments
Beryllium	Beryllium is found in beryllium base metal, beryllium alloys, and some filler metals. Breathing even small amounts of dust containing beryllium or beryllium fumes can cause lung inflammation, serious lung disease, and cancer.
Bismuth	
Cadmium	Cadmium is found in cadmium-coated base metals, filler metals, and fluxes. The fumes are extremely toxic.
Cleaning compounds and solvents	Cleaning compounds and solvents are potentially hazardous to the health. Gasoline and benzene, for example, produce toxic fumes and are also flammable. The vapors of some solvents, such as trichloroethylene, create toxic halogens and phosgene when in contact with a hot object or a welding arc.
Columbium (Niobium)	Columbium (niobium) contains many highly toxic compounds. Metallic columbium dust is an eye and skin irritant and also can be a fire hazard.
Lead	Lead, lead alloys, and lead coatings or paints containing lead can be dangerous. Breathing lead fumes can cause a wide variety of serious health problems, ranging from simple eye irritation to damage to the kidneys, heart, liver, and brain.
Manganese	Manganese is found in many welding products, such as electrodes and welding rods. Manganese is toxic to the brain and central nervous system when the levels in the body exceed normal limits. Breathing welding fumes containing manganese over a long period of time can lead to *manganism*, a condition similar to Parkinson's disease.

(continued)

Table 1-2 (continued)

Type	Comments
Mercury	Mercury is found in some paints and coatings used on metals. Breathing mercury vapors can result in serious pulmonary and neurological disorders. Symptoms include coughing, chest pains, nausea, vomiting, diarrhea, fever, and a metallic taste in the mouth.
Zinc	Zinc oxide fumes are released when the zinc layer on zinc-coated metals is melted. These fumes result in a condition known as *metal fume fever* or *zinc chills*, with symptoms resembling a viral flu.
Zirconium	Zirconium is found in beryllium base metal and alloys. Breathing zirconium fumes can cause irritation to the respiratory tract. Symptoms may include coughing, shortness of breath, sore throat, and runny nose.

when working with zinc, lead, or cadmium, can be toxic. These conditions also hold true for braze welding, brazing, and soldering.

Other sources of hazardous fumes produced during the various welding, joining, and cutting processes include antimony, bismuth, chromium, cobalt, copper, magnesium, molybdenum, nickel, thorium, and vanadium.

In the case of permanent welding stations, the size of the working area is an important consideration. Overcrowding will reduce the effectiveness of any ventilating system. The ventilating system itself (exhaust fans, etc.) should be designed to keep the level of toxic fumes and other contaminants at or below the maximum permitted level. Individual respiratory equipment is sometimes necessary when room ventilation is inadequate.

Protective Clothing

The manufacturers of welding equipment and supplies offer a broad range of protective clothing for welders, such as leather jackets; cape sleeves with detachable bib; waist, bib, or split-leg aprons; and shirt sleeves with snap fasteners at the wrist and adjustable leather straps at the top of the arm.

- Pants, shirts, and other clothing should be made of a flame-resistant material. The pants should be without cuffs, because cuffs can trap sparks and bits of molten metal.

- The clothing must be thick enough to minimize or prevent penetration by the dangerous radiation given off by an arc. The arc rays produce very strong visible and invisible rays (both ultraviolet and infrared) that can burn the eyes and skin. This radiation cannot be seen, but it is present. Any exposed skin can be burned quickly by these rays, which cause skin burns similar to sunburn.

- Sleeve cuffs should be tight against the wrist to prevent trapping flying sparks or molten particles. For this purpose, elastic bands or gauntlet cuffs are recommended.

- Black, flame-resistant cotton twill is often recommended for use with inert gas arc welding. Protective clothing made from this type of cloth is cheaper and lighter than leather.

- Wear high-top leather shoes, work shoes, or boots. Tennis shoes are not acceptable footwear.

2. OXYACETYLENE WELDING

Oxyacetylene welding (OAW) is a welding process in which the heat for welding is produced by burning a mixture of oxygen and acetylene. It is commonly referred to as *gas welding*.

NOTE

Oxyfuel welding (OFW) is the American Welding Society's name for any welding process in which coalescence is produced by a gas flame obtained by combining oxygen and an appropriate fuel gas. Oxyacetylene welding is the most widely used gas welding process in the oxyfuel group, and it is the one that will be covered in this section of the *Welding Pocket Reference*.

OXYACETYLENE WELDING APPLICATIONS

Oxyacetylene was the first welding process used commercially and industrially, especially for welding cast iron, wrought iron, low-alloy steels, copper, and bronze. Except for repair and maintenance work, oxyacetylene welding has been replaced by various arc welding processes, such as shielded metal arc welding (or stick welding), gas metal arc welding (or MIG welding), and tungsten metal arc welding (or TIG welding). Although relegated to a minor role in welding, oxyacetylene is still widely used for a wide variety of nonwelding uses such as cutting, preheating and post-heating, flame hardening, case hardening, braze welding, brazing, soldering, and descaling.

Oxyacetylene Welding (OAW) Advantages and Disadvantages

OAW Advantages:

- Self-contained and easily portable equipment
- Widely available equipment
- Relatively inexpensive equipment
- Easy to learn

OAW Disadvantages:

- Slower welding process than others.
- Uses volatile and potentially dangerous gases.
- Fuel gas and oxygen cylinders require special handling to avoid damage. Damaged cylinders can cause fire or explosions.

OXYACETYLENE WELDING EQUIPMENT

A typical oxyacetylene welding station will include the following components: (1) welding torch and nozzle; (2) oxygen cylinder, oxygen regulator, and oxygen hose; (3) acetylene cylinder, acetylene supply, acetylene regulator, and acetylene hose; (4) flashback arrestors and check valves; and torch lighter/sparklighter (see Figure 2-1).

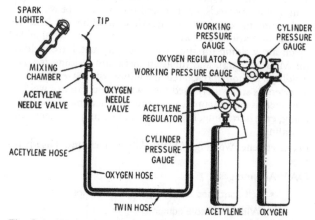

Fig. 2-1 Principal components of a typical oxyacetylene welding station.

Welding Torch

The welding torch is designed to mix oxygen and acetylene in nearly equal amounts, and then ignite and burn the gas mixture at the torch tip. The welding torch has two tubes (one for oxygen and the other for acetylene), a mixing chamber, and oxygen and acetylene valves to control and adjust the flame (see Figure 2-2).

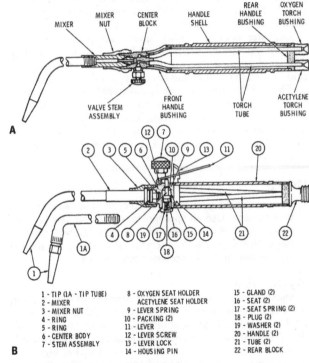

Fig. 2-2 Typical oxyacetylene torch.

1 - TIP (1A - TIP TUBE)
2 - MIXER
3 - MIXER NUT
4 - RING
5 - RING
6 - CENTER BODY
7 - STEM ASSEMBLY
8 - OXYGEN SEAT HOLDER
 ACETYLENE SEAT HOLDER
9 - LEVER SPRING
10 - PACKING (2)
11 - LEVER
12 - LEVER SCREW
13 - LEVER LOCK
14 - HOUSING PIN
15 - GLAND (2)
16 - SEAT (2)
17 - SEAT SPRING (2)
18 - PLUG (2)
19 - WASHER (2)
20 - HANDLE (2)
21 - TUBE (2)
22 - REAR BLOCK

Torch Tips

Welding tips may be purchased in a wide variety of sizes and shapes. The suitability of a particular welding tip design depends on a number of factors, including the accessibility of the area being welded, the rate of welding speed desired, and the size of the welding flame required for the job.

Manufacturers have their own numbering systems for indicating the different welding tip sizes. There is no industry standard, although there are comparison charts available. Tip size identifications have no bearing on minimum or maximum gas consumption or flame characteristics.

Drill size alone also fails to give an adequate comparison between the various makes of welding tips with identical tip drill sizes, because internal torch and tip construction may cause gas exit velocities and gas pressure adjustments to vary.

The welding torch nozzle is replaceable and is available in a wide variety of sizes. The size selected will depend on the thickness of the metal being welded. Table 2-1 contains data for the selection of welding tips. These are recommended sizes, and all variables should be taken into consideration before making the final selection.

NOTE

Because tips are subject to wear, they must be replaced from time to time. Use an appropriate wrench for this purpose (never pliers). Malfunctions such as *backfire*, *blowback*, and *popping out* will be greatly (if not completely) reduced by using the appropriate tip at the recommended pressure.

Oxygen Cylinders

Oxygen cylinders are seamless steel containers holding about 244 cubic feet of oxygen at a pressure of 2200 psi at 70°F. Smaller cylinders holding about 122 cubic feet is also available. A typical oxygen cylinder (see Figure 2-3) has an outside diameter of approximately 9 inches, is 54 inches high, and weighs (empty) between 104 and 139 pounds. The difference

Table 2-1 Oxyacetylene Welding Tip Data

Tip Size	Drill Size	Oxygen Pressure (psi) Min.	Oxygen Pressure (psi) Max.	Acetylene Pressure (psi) Min.	Acetylene Pressure (psi) Max.	Acetylene Consumption (CFH*) Min.	Acetylene Consumption (CFH*) Max.	Metal Thickness
000	75	1/4	2	1/2	2	1/2	3	up to 1/32"
00	70	1	2	1	2	1	4	1/64"–3/64"
0	65	1	3	1	3	2	6	1/32"–5/64"
1	60	1	4	1	4	4	8	3/64"–3/32"
2	56	2	5	2	5	7	13	1/16"–1/8"
3	53	3	7	3	7	8	36	1/8"–3/16"
4	49	4	10	4	10	10	41	3/16"–1/4"
5	43	5	12	5	15	15	59	1/4"–1/2"
6	36	6	14	6	15	55	127	1/2"–3/4"
7	30	7	16	7	15	78	152	3/4"–11/4"
8	29	9	19	8	15	81	160	11/4"–2"
9	28	10	20	9	15	90	166	2"–21/2"
10	27	11	22	10	15	100	169	21/2"–3"
11	26	13	24	11	15	106	175	3"–31/2"
12	25	14	28	12	15	111	211	31/2"–4"

*Oxygen consumption is 1.1 times acetylene consumption under neutral flame conditions. Gas consumption data is merely for rough estimating purposes. It will vary greatly with the material being welded and the particular skill of the operator. Pressures are approximate for hose lengths up to 25 feet. For longer hose lengths, increase pressures about 1 psi per 25 feet of hose.

(Courtesy Victor Equipment Co.)

17

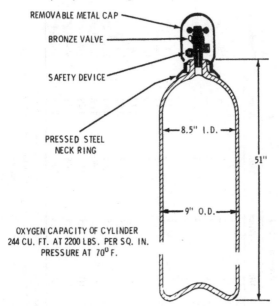

REMOVABLE METAL CAP

BRONZE VALVE

SAFETY DEVICE

PRESSED STEEL
NECK RING

8.5" I.D.

51"

9" O.D.

OXYGEN CAPACITY OF CYLINDER
244 CU. FT. AT 2200 LBS. PER SQ. IN.
PRESSURE AT 70° F.

Fig. 2-3 Typical oxygen cylinder. *(Courtesy Airco Welding Products)*

in weight depends on the type of steel used to construct the cylinder. A full cylinder will increase the weight by about 20 pounds.

The oxygen supplied through these cylinders is about 99.5% pure. The oxygen pressure will vary according to temperature changes, but the weight and percentage of oxygen remain the same.

A special bronze valve is inserted into the top of the oxygen cylinder, and a removable cap is placed over it for protection. The oxygen cylinder valve is designed to withstand great pressures. If the pressure becomes too great, a disc in the safety plug located on the oxygen valve will break, allowing the excess oxygen to escape before the high pressures rupture the cylinder. Pressure regulators can be attached to the oxygen cylinder valve at an outlet that has a standard male thread.

Variations in oxygen cylinder pressures with respect to temperature changes are given in Table 2-2. Oxygen cylinder content, as indicated by gauge pressure at 70°F (for a 244-cubic-foot cylinder), is listed in Table 2-3.

Table 2-2 **Variations in Oxygen Cylinder Pressures with Temperature Changes**

Temperature Degrees F	Pressure, psi (approx.)	Temperature Degrees F	Pressure, psi (approx.)
120	2500	30	1960
100	2380	20	1900
80	2260	10	1840
70	2200	0	1780
60	2140	−10	1720
50	2080	−20	1660
40	2020		

Gauge pressures are indicated for varying temperature conditions on a full cylinder initially charged to 2200 psi at 70°F. Values are identical for 244-cubic-foot and 122-cubic-foot cylinders
(Courtesy Airco Welding Products)

CAUTION

Because oil or grease in contact with oxygen may cause a violent and explosive reaction resulting in serious injuries or even death, the following precautions must be taken:

- Post a warning note on or near the oxygen cylinder that forbids the use of oil or grease on the cylinder fittings. Always keep oxygen cylinders and apparatus free from oil, grease, and other flammable or explosive substances.

Table 2-3 Oxygen Cylinder Content*

Gauge Pressure (psi)	Content (Cu. Ft.)	Gauge Pressure (psi)	Content (Cu. Ft.)
190	20	1285	140
285	30	1375	150
380	40	1465	160
475	50	1550	170
565	60	1640	180
655	70	1730	190
745	80	1820	200
840	90	1910	210
930	100	2000	220
1020	110	2090	230
1110	120	2200	244
1200	130		

* 122-cubic-foot cylinder content one-half above volumes.
(Courtesy Airco Welding Products)

- Never handle oxygen cylinders or equipment with oily hands or gloves.

Acetylene Cylinders

Acetylene cylinders are available in sizes ranging from a capacity of 10 cubic feet up to 360 cubic feet. Cylinders are available on lease or are sold on an exchange basis. Figure 2-4 illustrates the construction of a typical acetylene cylinder. Note the removable cap cover over the valve at the top.

Acetylene can be used safely if is dissolved under a pressure of 250 psi in liquid acetone, which is used to stabilize the acetylene. Always secure an acetylene cylinder in an upright position on a hand truck or against a wall to prevent acetone leakage. Never drop or strike an acetylene cylinder. Even a slight shock can cause a cylinder to explode. Acetylene is very unstable at elevated pressures.

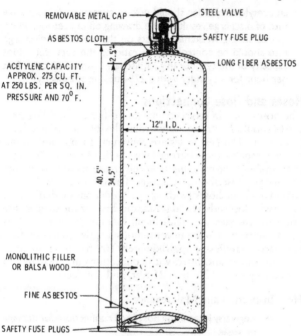

REMOVABLE METAL CAP — STEEL VALVE
ASBESTOS CLOTH — SAFETY FUSE PLUG
2.5"
ACETYLENE CAPACITY — LONG FIBER ASBESTOS
APPROX. 275 CU. FT.
AT 250 LBS. PER SQ. IN.
PRESSURE AND 70° F.

12" I.D.

40.5"
34.5"

MONOLITHIC FILLER
OR BALSA WOOD

FINE ASBESTOS

SAFETY FUSE PLUGS

Fig. 2-4 Acetylene cylinder. *(Courtesy Airco Welding Products)*

NOTE

A damaged cylinder must be removed immediately to the outside of the building. No chances should be taken with a leaking acetylene gas cylinder.

CAUTION

Never exceed a 15-psi regulator gauge reading when reducing the acetylene to a working pressure. The discharge rate of an acetylene cylinder should not be excessive. An excessive rate of discharge results in the drawing off of acetone, which causes strong weakening effects on the weld. The discharge rate should be controlled as a ratio to the total cubic-foot capacity of the cylinder (e.g., a discharge rate of 50 cubic feet per hour for a cylinder with a 275-cubic-foot capacity).

Hoses and Hose Connections

The hoses used in oxyacetylene and other gas welding processes should be strong, nonporous, flexible, and not subject to kinking. The best hoses are constructed from nonblooming neoprene tubing reinforced with braided rayon. The outer coating should be resistant to oil and grease and tough enough to survive most shop conditions. These hoses may be purchased as single hoses (with $\frac{1}{8}$ - to $\frac{1}{2}$-inch inside diameters), or as twin double-barreled hoses (one line for the oxygen, the other for the fuel gas) with metal binders at the base. Gas hoses can be purchased in continuous lengths up to 300 feet. Standard reel lots are approximately 100 feet long. In addition, hoses cut and fitted with connections are available in boxed lengths of $12\frac{1}{2}$ feet, 25 feet, and 100 feet.

Hose Inspection and Maintenance

- Regularly inspect the hoses and connections for damage or wear.
- Replace leaky, worn, or damaged hoses. Do not attempt to repair them. A tape repair on a hose is not a safe seal.
- Replace damaged hose connections.
- Always blow out the hoses before welding or cutting.
- Keep as much hose off the floor as possible to protect it from being run over by equipment or stepped on.
- Keep hose runs as short as possible to avoid damage.
- Coil excess hose to avoid kinking and tangles in the line.

The oxygen hose is usually black or green (blue in the United Kingdom). The fuel gas hose is generally red. These color distinctions are for safety reasons. For example, using an oxygen hose to carry acetylene could cause a serious accident. No hose should be used for any purpose other than its initial use. A further safety precaution is the design of the threading on the hose connections. The fuel gas hoses have left-hand threaded connections (and a groove around the outside), whereas the oxygen hoses are fitted with right-hand thread connections and swivel nuts at both ends.

NOTE

Always use the shortest possible hose length between the cylinders and the torch. You will use less oxygen and acetylene with shorter hoses than with longer ones. You will also experience less pressure drop at the torch when using shorter hoses.

A metal clamp is used to attach the welding hose to a nipple. A nut on the other end of the nipple is connected to the regulator or torch. Sometimes the shape of the nipple will indicate its use. A bullet-shaped nipple is generally used for oxygen hoses and a nipple with a straight taper is used for fuel gas hoses. Another identification method is a groove running around the center of the acetylene nut. This is an indication of a left-hand thread that will only screw into the acetylene outlet.

NOTE

Multiple sets of hoses may be connected to a single regulator on a single set of oxyacetylene cylinders only by installing an approved, commercially available fitting listed by a nationally recognized testing laboratory. This fitting must be installed on the output side of the regulator and must have an integral shutoff and reverse-flow check valve on each branch.

Pressure Regulators

The oxygen and acetylene pressure regulators are used to control gas pressure. They do this by reducing the high pressure of the gases stored in the cylinders to a working pressure delivered to the torch, and by maintaining a constant gas working pressure during the welding process. There is a regulator for the oxygen and another for the fuel gas. These gas pressure regulators are connected between the gas cylinder and the hose leading to the torch.

Regulators may differ according to their capacity (ranging from light to high) and according to the type of gas for which they are designed. For example, an oxygen regulator *cannot* be used for acetylene gas, and vice versa.

Oxygen regulators do not have the same thread size as acetylene regulators. This is a safety feature that prevents their being attached to the wrong cylinder. Another safety feature is found in the hose outlet from the regulator. The outlet connections for oxygen hoses have right-hand threads. For acetylene and other fuel gases the outlet connections have left-hand threads.

Both single-stage and two-stage regulators are available for use in welding systems. Many regulators are constructed with two gauges. One gauge (the high-pressure gauge) indicates the pressure of the gas in the cylinder, and the other (the low-pressure gauge) indicates the working pressure of the gas being delivered to the torch.

A single-stage regulator requires torch adjustments to maintain a constant working pressure. As the cylinder pressure falls, the regulator pressure also falls, necessitating torch adjustment. In the two-stage regulator, there is automatic compensation for any drop in cylinder pressure. The two-stage regulator is virtually two regulators in one that operate to reduce the pressure progressively in two stages. The first stage in the two-stage regulator serves as a high-pressure reduction chamber. A predetermined pressure is set and maintained by a spring and diaphragm. The gas then flows at reduced pressure into the second of the two stages, which serves as a low-pressure reduction chamber. In this second chamber, pressure is controlled by an adjustment screw.

Regulator Maintenance

Proper regulator maintenance is important for safe and efficient operation. The following points are particularly important to remember:

- The adjusting screw on the regulator must *always* be released before opening the cylinder valve. Failure to do this results in extreme pressure against the gauge that measures the line pressure, and it may damage the regulator.

- Gauge-equipped regulators should never be dropped, improperly stored, or be subjected to careless handling in any way. The gauges are extremely sensitive instruments and can easily be rendered inoperable.

- *Never* oil a regulator. Most regulators will have the instruction "use no oil" printed on the face of both gauges.

- All regulator connections should be tight and free from leaks.

CAUTION

Never adjust an acetylene regulator to allow a discharge greater than 15 psi (103.4 kPa) gauge.

Flashback Arrestors and Check Valves

Flashback arrestors, also called *flame traps*, and spring-loaded nonreturn check valves should be installed between the acetylene and oxygen openings in the torch and the matching hoses.

NOTE

Some torches are designed with integral flashback arrestors and check valves.

Flashback

Flashback is a potentially dangerous condition caused by the burning of an oxygen and fuel gas mixture in the mixing chamber of the torch handle instead of at the torch tip. If the burning fuel gas-and-oxygen mixture passes through the hoses and regulators and into the cylinders, it can cause a fire or explosion resulting in serious injury, or even death. Among the causes of flashback are:

- Opening the fuel gas and oxygen cylinder valves and then attempting to light a torch with a blocked tip
- Loose hose connections and/or hose leaks
- Low gas velocity produced by incorrect gas pressure
- Lighting the torch with a failed oxygen or acetylene regulator

The hose check valves prevent the oxygen and fuel gas from crossing over and mixing together in a volatile mixture at the back of the torch mixing chamber. The flame arrestors, which commonly are narrow stainless steel tubes, stop the flame by absorbing its heat and constricting its passage.

Sparklighters

The oxygen and fuel mixture should not be ignited with a match. A sudden flareup on ignition could cause the welder's hands or other parts of the body to be burned. To insure against such danger, the gas mixture should be ignited with a device that provides the required degree of safety for the welder. One such device is a sparklighter—a simple, inexpensive device using flint and steel. Some sparklighters are equipped with pistol grips, and they shoot a shower of sparks at the gas flowing from the torch. Others have a rotating flint holder that permits longer use before having to insert a new flint.

GAS WELDING RODS

Gas welding rods, or filler rods, are small-diameter metal rods used to add metal to the weld during the welding process. During welding, the filler rod melts and deposits its metal into the puddle, where it joins with the molten base metal to form a strong weld. Because the composition of the filler rod must be matched as closely as possible to that of the base metal, the selection of the appropriate rod for the job is extremely important. Choosing the wrong filler rod will result in a weak and ineffective weld.

Welding rods are available in a variety of sizes and compositions (see Table 2-4). The sizes range in diameter from 1/16 inch to 3/8 inch. Cast-iron rods are sold in 24-inch lengths, and all others are available in 36-inch lengths.

Table 2-4   Gas Welding Filler Rods

Type	Comments
RG45	• Copper-coated low-carbon steel rod for gas welding.
	• AWS A5.2 Class RG45.
	• Use a neutral flame to avoid excess oxidation or carbon pickup.
	• Commonly used to weld ordinary low-carbon steel up to 1/4″ thick.
	• Recommended where ductility and machinability are most important. Produces high-quality welds that are ductile and free of porosity.
	• Excellent for steel sheet, plates, pipes, castings, and structural shapes. No flux required.

(continued)

Table 2-4 (continued)

Type	Comments
RG60	• Low-alloy steel gas welding rod.
	• AWS A5.2 Class RG60.
	• Use a neutral flame to avoid excess oxidation or carbon pickup.
	• Used to produce quality welds with high tensile strength on low-carbon and low-alloy steels such as sheets, plates, pipes of grades A and B analysis, and structural shapes.
	• Recommended for critical welds that must respond to the same annealing and heat treatment as regular grades of cast steel.
	• The high silicon and manganese composition removes impurities from the molten metal, thereby eliminating the need for flux.
	• RG60 rod is also used as a filler metal in gas tungsten arc welding (GTAW/TIG).
Bare brass rod	• Low-fuming bronze gas welding rods made of copper-tin alloy that flows easily and joins a variety of metals, including cast and malleable iron, galvanized steel, brass, copper, and steel.
	• Produces strong joints up to 63,000 psi tensile strength.
	• Flux is required for bare brass rod.
Flux-coated brass or low-fuming bronze rod	• Same characteristics as bare brass rod.
	• Flux is contained in the rod coating.

FLUX

A flux is a material used to prevent, dissolve, or facilitate the removal of oxides and other undesirable substances that can contaminate the weld. The flux material is fusible and nonmetallic. The chemical reaction between the flux and the oxide forms a slag that floats to the top of the molten puddle of metal during the welding process. The slag can be removed from the surface after the weld has cooled. Note:

- A flux is selected for its chemical composition, depending on the metal or metals to which it is applied. Fluxes may be divided into three main categories: (1) welding fluxes, (2) brazing fluxes, and (3) soldering fluxes. Welding fluxes are categorized as gas welding fluxes and braze welding fluxes.

- Fluxes are sold as powders, pastes, or liquids (frequently in plastic squeeze bottles). Quantities are available in 1/2- to 5-pound cans or jars, or in drums of 25 pounds or more.

- Some powdered fluxes may be applied by dipping the heated welding rod into the can. The flux will stick to the rod. The flux powder may also be applied directly to the surface of the base metal. Some powdered fluxes are mixed with alcohol or water and applied to the surface as a paste.

- Paste or liquid fluxes are applied in their purchased form.

WELDING TIP CLEANERS

Welding torch tips must be cleaned regularly in order to prolong the life of the tip and to provide consistently high performance. Stainless steel tip cleaners (wires) are available in various diameters to fit different tip openings (orifices). All deposits on the inside of the tip must be removed without enlarging the size of the orifice.

Oxyacetylene Welding Equipment Setup

1. Fasten the oxygen and acetylene cylinders in an upright position to a welding cart, wall, or a fixed vertical surface to keep them from falling over. Use a chain or other nonflammable material to fasten them in place. Locate them as close as possible to the welding job, but away from open flames.

2. Remove the caps from both cylinders. Examine the cylinder outlet nozzles for stripped threads or a damaged connection seat.

3. Open and close the oxygen cylinder valve very quickly to blow out any loose dust or dirt that may have accumulated in the outlet nozzle. Wipe the connection seat with a dry, clean cloth. If the dust or dirt is not removed, it could damage the regulator and cause incorrect gauge readings.

CAUTION

Turn your head away when opening the oxygen cylinder valve, and make sure the oxygen stream is not directed toward another worker, a spark, or an open flame. The high pressure of the oxygen stream can cause serious injury to the eyes. Oxygen can't be ignited; rather, it *facilitates* burning and can make a spark or flame much more intense.

NOTE

The valve opening and the inlet nipple should be shiny and clean, both inside and outside. This is particularly important for the oxygen cylinder. Oil or grease in the presence of oxygen is flammable, or even explosive. Never allow oxygen to contact oil, grease, or other flammable substances.

4. Repeat Step 3 for the acetylene cylinder.

5. Connect the oxygen regulator to the oxygen cylinder.

NOTE

Regulators must be used only with the gas and pressure range for which they are intended and marked. Cylinder valve outlets and the matching inlet connections on regulators have been designed to minimize the chances of making incorrect connections.

6. Connect the acetylene regulator to the acetylene cylinder.

7. Connect the green or black oxygen hose to the oxygen regulator. The oxygen hose has a right-hand thread and must be turned clockwise to tighten. Make the connection tight, but avoid overtightening it.

8. Connect the red acetylene hose to the acetylene regulator. The acetylene hose has a left-hand thread and must be turned counterclockwise to tighten. Again, do not overtighten the connection.

9. Charge the oxygen regulator by slowly opening the oxygen cylinder valve. Opening the valve slowly prevents damage to the regulator seat.

CAUTION

Never face the regulator when opening the cylinder valve. A defective regulator may allow the gas to blow through with enough force to break the gauge glass, resulting in possible injury to anyone standing nearby. Always stand to one side of the regulator when opening the cylinder valve, and turn the valve slowly.

10. Open the oxygen regulator adjusting screw (the T-handle on the regulator), blow out any dirt or debris in the oxygen hose, and then close the screw.

11. Repeat Steps 9 and 10 for the acetylene regulator and hose.

12. Connect the oxygen hose to the oxygen needle valve on the torch, and connect the acetylene hose to the acetylene needle valve. Again, the oxygen hose has a right-hand thread, whereas the acetylene hose has a left-hand one.

13. Close the welding torch needle valves and open the oxygen and acetylene cylinder valves. Adjust the regulators for a normal working pressure and check for leaks at all the connections, using soap and water. Soap bubbles will indicate a leak. Tighten the connections with a wrench. If this fails to eliminate the problem, shut off the oxygen and acetylene and check for stripped threads, defective hoses (old hoses become porous), or other damaged parts, and repair or replace them as necessary.

NOTE

Leaks must be repaired before attempting to use the welding equipment. A leak not only results in wasted gas, but it can also cause a fire or an explosion. Always check the equipment for leaks regularly, not just when the equipment is initially set up.

Welding Tip Selection

The tip size will depend on the thickness of the metal being welded. Use a tip with a small opening for welding thin sheet metal.

NOTE

It is very important to use the correct tip size for the proper working pressure. If too small a tip is employed, the heat will not be sufficient to fuse the metal to the proper depth. When the tip is too large, the heat is too great and can burn holes in the metal.

1. Select the correct tip size for the welding job. The size of the welding tip depends on many factors, including the thickness of the metal, the welding position, and the type of metal being welded. If the tip is too small for the work, too much time is wasted in making the weld and poor fusion is likely to result. Too large a tip is likely to produce poor metal in the weld (caused by overoxidation) and a rough-looking job resulting from lack of control of the flowing metal.

2. Install the tip in the nozzle.

Lighting the Torch

1. Point the torch tip down and away from your body.

2. Open the oxygen and acetylene cylinder valves and set the working pressure to correspond to the size of tip being used.

CAUTION

Always face away from the regulator when opening a cylinder valve. A defect in the regulator may cause the gas to blow through, shattering the glass and blowing it into your face. Remember, oxygen and acetylene are under high pressure in the tanks, and if the gas is forced against the regulator suddenly, it may cause some damage to the equipment.

3. Open acetylene cylinder valve approximately one complete turn and the oxygen all the way. Next turn the oxygen and acetylene regulator adjusting valves to the required working pressures.

4. Open the acetylene needle valve on the torch about one-quarter turn and spark the lighter at the torch tip.

CAUTION

If you take too long to spark the lighter, acetylene will build up around the torch tip. When the excess acetylene is finally lit, it may cause an explosion resulting in burns to the hand.

NOTE

Never use a match to light a torch. It brings your fingers too close to the tip, and the sudden ignition of the acetylene can cause serious burns.

CAUTION

Make no attempt to relight a torch from the hot metal when welding in an enclosed box, tank, drum, or other small cavity. There may be just enough unburned gas in this confined space to cause an explosion as the acetylene from the tip comes in contact with the hot metal. Instead, move the torch to the open, relight it in the usual manner, and make the necessary adjustments before resuming the weld.

5. After igniting the acetylene gas, make adjustments at the acetylene valve to give the proper intensity of flame. If the acetylene flame is accompanied by a lot of smoke, increase the amount of acetylene until the smoke disappears and the flame seems to jump off the torch tip.

6. When the acetylene flame is properly adjusted, open the oxygen valve slowly so that air in the line escapes gradually and does not blow out the flame.

7. The torch oxygen valve is then gradually opened until the flame changes from a ragged yellow flame to a perfectly formed bluish cone. This flame, known as a *neutral* flame, is the torch flame commonly used for most welding.

Flame Adjustment

The proportions of oxygen and acetylene can be adjusted to produce a neutral, oxidizing, or carburizing flame. Oxyacetylene welding is normally performed using a neutral flame produced by mixing roughly equal amounts of oxygen and acetylene: See Tables 2-5 and 2-6, and Figure 2-5.

Table 2-5 Basic Components of the Oxyacetylene Flame

Flame Components (see Figure 2-5)	Comments
Inner tip	Also called the *cone*, it is that portion of the flame located at the bore of the torch nozzle; the innermost portion of the flame. Caution: Never allow the inner tip of the flame to touch the work, because it will burn through the metal.
Beard, or brush	Located between the outer envelope and the inner tip, or cone, of the flame. It is not always present. Caution: Never allow the beard, or brush, to touch the work.
Outer envelope	Extends around the beard and inner tip. It is much larger in volume than the beard and inner tip because it is fed oxygen from the surrounding atmosphere.

Table 2-6 Oxyacetylene Flame Types

Flame Type (see Figure 2-6)	Comments
Acetylene flame	Very white, large, smoky flame produced when torch is first lit.
Carburizing flame	Also called a *reducing flame*, produced by burning an excess of acetylene. The flame very often has no beard or brush on its inner tip. When adjusted to have a small beard, it may be used on most nonferrous metals (those not containing iron elements). The outer envelope is usually the portion of the flame used on these metals.

(*continued*)

Table 2-6 *(continued)*

Flame Type (see Figure 2-6)	Comments
Neutral or flame	Produced by burning one part acetylene gas and slightly more than one part oxygen.
Oxidizing flame	Produced by burning an excess of oxygen. The flame has no beard, and both the inner tip and the envelope are shorter. The oxidizing flame is of limited use because it is harmful to many metals.

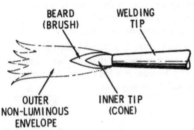

Fig. 2-5 Three basic components of the oxyacetylene flame.

Shutting Off the Torch

1. First, close the acetylene needle valve on the torch, because shutting off the flow of acetylene will immediately extinguish the flame. If the oxygen is shut off first, on the other hand, the acetylene will continue to burn, throwing off smoke and soot.

2. Close the oxygen needle valve on the torch.

Closing Down the Equipment

1. If the entire welding unit is to be shut down, shut off both the acetylene and the oxygen cylinder valves.

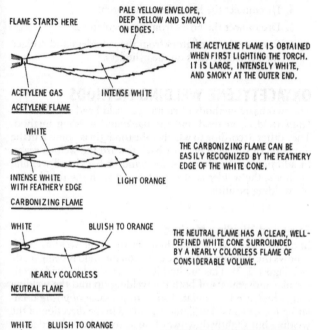

FLAME STARTS HERE

PALE YELLOW ENVELOPE, DEEP YELLOW AND SMOKY ON EDGES.

THE ACETYLENE FLAME IS OBTAINED WHEN FIRST LIGHTING THE TORCH. IT IS LARGE, INTENSELY WHITE, AND SMOKY AT THE OUTER END.

ACETYLENE GAS   INTENSE WHITE

ACETYLENE FLAME

WHITE

INTENSE WHITE WITH FEATHERY EDGE   LIGHT ORANGE

CARBONIZING FLAME

THE CARBONIZING FLAME CAN BE EASILY RECOGNIZED BY THE FEATHERY EDGE OF THE WHITE CONE.

WHITE   BLUISH TO ORANGE

NEARLY COLORLESS

NEUTRAL FLAME

THE NEUTRAL FLAME HAS A CLEAR, WELL-DEFINED WHITE CONE SURROUNDED BY A NEARLY COLORLESS FLAME OF CONSIDERABLE VOLUME.

WHITE   BLUISH TO ORANGE

NEARLY COLORLESS

OXIDIZING FLAME

THE OXIDIZING FLAME CAN BE RECOGNIZED BY ITS SHORTER ENVELOPE OF FLAME AND THE SMALL POINTED WHITE CONE.

Fig. 2-6 Oxyacetylene flame types.

2. Remove the pressure on the working gauges by opening the needle valves until the lines are drained. Then promptly close the needle valves.

3. Release the adjusting screws on the pressure regulators by turning them to the left.

4. Disconnect the hoses from the torch.

5. Disconnect the hoses from the regulators.

6. Remove the regulators from the cylinders and replace the protective caps on the cylinders.

OXYACETYLENE WELDING METHODS

The two basic methods of running a weld bead are the *forehand* welding method and the *backhand* welding method. They differ according to whether the torch tip is pointed in the direction of the weld bead or back toward the welded seam. In addition to the choice of forehand or backhand welding methods, the welder is also confronted with the problem of the welding position.

Forehand Welding Method

In the forehand welding method, the tip of the welding torch follows the welding rod in the direction the weld is being made (see Figure 2-7). This method is characterized by wide semicircular movements of both the welding tip and the welding rod, which are manipulated so as to produce *opposite* oscillating movements. The flame is pointed in the direction of the welding but slightly downward, so as to preheat the edges of the joint.

The major difficulty with the forehand welding method is encountered when welding thicker metals. In order to obtain adequate penetration and proper fusion of the groove surfaces, as well as to permit the movements of the tip and rod, a wide V-groove (90° included angle) must be created at the joint. This results in a large puddle that can prove difficult to control, particularly in the overhead position.

NOTE

Forehand welding is also sometimes referred to as *ripple welding* or *puddle welding*.

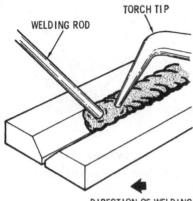

TORCH TIP
WELDING ROD
DIRECTION OF WELDING

Fig. 2-7 Forehand welding method: The welding rod moves ahead of the torch tip.

Backhand Welding Method

In the backhand welding method, the tip of the torch *precedes* the welding rod in the direction the weld is being made (see Figure 2-8). In contrast to the forehand welding method, the flame is pointed back at the puddle and the welding rod. In addition, the torch is moved steadily down the groove without any oscillating movements. The welding rod, on the other hand, may be moved in circles (within the puddle) or semicircles (back and forth around the puddle).

Backhand welding results in the formation of smaller puddles. A narrower V-groove (30° bevel, or 60° included angle) is required than with the forehand welding method. As a result, backhand welding provides greater control and reduced welding costs.

Adding Filler Metal

The end of the welding rod should be melted by keeping it beneath the surface of the molten weld puddle. *Never* allow

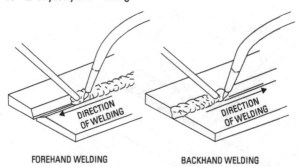

FOREHAND WELDING   BACKHAND WELDING

Fig. 2-8 Backhand welding method: The torch tip moves ahead of the welding rod.

it to come into contact with the inner cone of the torch flame. *Do not* hold the welding rod *above* the puddle so that the filler metal drips into the puddle.

Torch Movement and Angle

Figures 2-9 through 2-13 illustrate (1) common torch movements, (2) recommended torch angle and distance of the

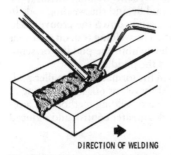

DIRECTION OF WELDING

Fig. 2-9 A bead produced by a varied welding torch speed. The narrow sections are produced by increasing the speed; at slower speeds, a larger puddle forms.

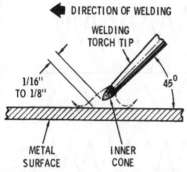

Fig. 2-10 Recommended torch angle and inner cone distance from the metal surface.

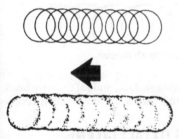

Fig. 2-11 A correctly formed weld bead, using an overlapping circular motion of the torch tip.

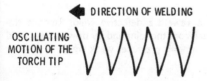

Fig. 2-12 An oscillating motion of the torch tip. The forward speed of the torch must be constant.

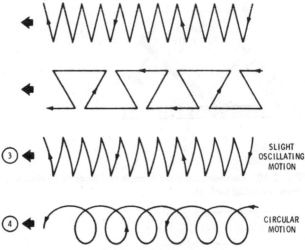

Fig. 2-13 Other examples of torch motion.

flame's inner cone tip from the surface, and (3) the weld bead produced by increasing and decreasing the welding speed.

TROUBLESHOOTING OXYACETYLENE WELDING

NOTE

Do not attempt to reweld a defective weld. Remove the defective weld metal from the joint and lay down a new weld bead. Rewelding commonly produces weak welds.

Table 2-7 Troubleshooting Oxyacetylene Welding

Problem	Possible Cause	Suggested Remedy
Welding flame splits.	Dirty or clogged torch tip.	Stop welding, allow the torch tip time to cool, thoroughly clean the tip, and then resume welding.
Sharp inner cone of flame disappears.	Dirty torch tip.	Stop welding, allow the torch tip time to cool, thoroughly clean the tip, and then resume welding.
Premature ignition of gas mixture.	Sparks from the weld puddle produce carbon deposits inside the nozzle and on the torch tip face.	Stop welding, allow the torch tip time to cool, thoroughly clean the tip, and then resume welding.
Backfire or popping sound (a single explosion or pop, or a series of small explosions occurring shortly after the torch is lit). The welding flame may disappear and then reappear, or remain extinguished.	a. Clogged torch tip. b. Preignition of gas mixture in the torch tip, torch mixing chamber, or both. c. Overheated torch tip caused by holding the tip too close to the work.	a. Stop welding, allow the torch tip to cool, clean the tip, and then resume welding. b. Increase oxygen and acetylene pressures slowly until the popping sound is eliminated. c. Stop welding, allow the torch tip time to cool, and then resume welding with the tip held farther away from the work.

(continued)

43

Table 2-7 (continued)

Problem	Possible Cause	Suggested Remedy
Squealing sound (resulting from a very rapid series of small explosions after the torch is lit).	Same as for backfire or popping sound.	Same as for backfire or popping sound.
Flashback (recession of flame into or in back of the torch mixing chamber).	Equipment problems such as loose or damaged torch tips, clogged torch tip orifices, or kinked or damaged hoses.	First close the torch oxygen valve, and then the acetylene valve. Wait to make sure the fire in the torch or hose has burned out.
	Incorrect welding procedures such as improper gas pressures, an overheated torch tip, or failure to purge torch hose lines before lighting the torch.	Check for damage to tips, regulators, and hoses, and replace them as necessary. Purge the hoses, check oxygen/acetylene pressures, and then relight the torch, using the standard lighting procedure.
Poor penetration.	Insufficient heat to penetrate to proper depth, because the torch tip is too small.	Replace the tip with one of appropriate size.
Holes burned in metal.	Excessive heat caused by a tip too large for job.	Replace the tip with one of appropriate size.

Problem	Cause	Remedy
Uneven weld (incomplete penetration; good penetration in spots and partial penetration in between).	The torch is moved along the joint too quickly or too slowly; zigzag movements not uniform nor in step with the puddle; zigzag movements overlapping.	Weld with a steady, uniform movement.
Fused portions at one side of the joint and not the other.	Caused by not moving the torch over the joint equally from side to side.	Move the torch so that it touches both sides of the joint uniformly.
Holes in the joint.	Caused by holding the flame too long in one place and overheating the metal.	Move the torch along the joint at a speed guaranteed not to overheat the base metal.
Holes in the joint at the end of the weld.	Failure to lift the torch and reduce the heat when reaching the end of the weld.	Use appropriate welding procedure.
Oxide inclusions (indicated by black specks on the broken surfaces of the weld).	Caused by not adequately cleaning the base metal surface prior to welding.	Make sure the base metal is as clean as possible before welding.

(continued)

Table 2-7 (continued)

Problem	Possible Cause	Suggested Remedy
Adhesions.	a. Insufficient heat was directed to one side of the joint. b. Welding speed too high, causing the weld metal to break cleanly from the metal surface on one side.	a. Apply heat uniformly across the joint. b. Decrease welding speed.
Brittle welds.	Caused by using a carburizing flame. It is important for the welder to use the correct flame when welding.	Use a neutral flame on ferrous metals.
Overheating.	Caused by moving the flame so slowly that too much heat is directed into the weld puddle. This results in the formation of excess metal *icicle* deposits on the bottom of the weld.	Increase torch speed.
Welding (filler) rod sticks in the weld puddle.	a. Failure to keep the weld puddle molten. b. Failure to dip the welding rod into the weld puddle fast enough.	a. Maintain heat over the weld puddle long enough to complete addition of the welding rod. b. Dip the welding rod into the puddle more quickly.

3. OXYACETYLENE CUTTING

Oxyacetylene cutting (OFC-A), sometimes called *flame cutting*, *oxygen cutting*, or *torch cutting*, is a thermal cutting process used to cut ferrous metals. The process involves first preheating the metal surface to its ignition temperature and then, from the torch tip, introducing a stream of pure oxygen under pressure to oxidize or rapidly burn the metal. The oxidizing metal produces a cut (kerf) behind the movement of the torch flame.

NOTE

The American Welding Society (AWS) employs the term *oxyfuel cutting* (OFC) to describe a group of cutting processes that use oxygen mixed with a fuel gas to heat the metal surface and a separate supply of oxygen to cut the metal. Oxyacetylene cutting, the most commonly used oxyfuel cutting process, combines oxygen with the fuel gas acetylene. The American Welding Society designation of this process is OFC-A.

OXYACETYLENE CUTTING APPLICATIONS

Manual oxyacetylene cutting is most commonly used in maintenance and repair work. Related processes based on the oxyfuel cutting process are flame gouging, scarfing, and flame machining.

Oxyacetylene Cutting Advantages and Disadvantages
Oxyacetylene Cutting Advantages:

- Self-contained, portable equipment.
- Easy to learn and use.
- Inexpensive equipment.
- Widely used to cut iron and steel.
- Castings and large plates may be cut in place without disassembly.

47

Oxyacetylene Cutting Disadvantages:

- Limited to cutting metals subject to oxidation (ferrous metals). Not recommended for cutting stainless steel, aluminum, and other nonoxidizing metals.

- Cuts are rough, leaving large amounts of slag that must be removed. Not recommended for jobs where precision cuts are required.

- Slower and more expensive than plasma arc cutting.

- Proper ventilation to control fumes is required.

OXYACETYLENE CUTTING EQUIPMENT

The same equipment is used for both oxyacetylene cutting and oxyacetylene welding, except for the torch. For cutting, the welding torch is replaced by either a cutting torch or a cutting attachment.

CAUTION

As a safety precaution, an oxyacetylene cutting system must be equipped with a reverse-flow check valve and flash arrestor in each hose, at the torch, and at the regulator. These devices will prevent any damage caused by the sustained burning back of the flame into the torch tip. A sustained burnback (flashback) can damage the hose and equipment.

Cutting Torch

The oxyacetylene one-piece cutting torch has two oxygen supply tubes and one fuel supply tube (see Figure 3-1). One oxygen supply tube provides the oxygen for mixing with the fuel gas to preheat the metal surface; the other oxygen supply tube supplies a separate oxygen flow to the nozzle to cut the metal.

The one-piece cutting head puts more distance between the cutting action and the welder. It can usually handle greater oxygen flows for large jobs. Some cutting torches have cutting heads at a particular angle for a given task, to relieve operator fatigue. The position of the cutting handle is a matter of preference and varies by manufacturer.

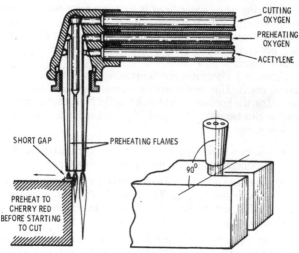

Fig. 3-1 Typical cutting torch showing the locations of the preheating oxygen tube, cutting oxygen tube, and acetylene (fuel gas) tube. *(Courtesy Airco Welding Products)*

Cutting Attachment

Hand cutting attachments have been developed to provide the standard welding torch with greater operating flexibility. These hand cutting attachments are connected to the body of the welding torch after removing the welding tip assembly.

Hand cutting torch attachments are available in models similar to standard cutting torches. These models include such design variables as: (1) the position of the oxygen control lever, (2) the angle of the cutting head (straight, 75°, and 90° angles), and (3) the fuel gas they use (acetylene or *MAPP*, natural gas or propane, etc.).

A conventional oxyacetylene welding outfit can easily be converted for oxyacetylene cutting by replacing the mixer and welding tip with a cutting attachment.

A cutting attachment head is less expensive than a one-piece cutting torch. It is also quicker and easier to change back and forth between the cutting and welding functions than between a welding torch and a cutting torch, with its greater length.

Figure 3-2 illustrates the construction of a typical hand cutting torch. This model has the cutting oxygen valve lever located on the bottom of the torch handle. Note that different head angles (straight, 45°, and 75°) are available. Also note the following:

- The two basic types of oxyacetylene cutting torches are the injector type and the equal pressure type.

- Acetylene is delivered to an injector-type torch at pressures below 1 psig. The gases in the equal-pressure torches, on the other hand, are delivered at pressures above 1 psig.

- An oxyacetylene cutting torch may be a torch designed specifically for cutting, or it may be the body of a welding torch with a hand cutting attachment fastened to it.

The procedure for connecting the hand cutting attachment to the body (handle) of a welding torch is as follows:

- Make certain the welding equipment is shut down, with all gases drained.

- Make certain the welding equipment is still connected and operational.

- Remove the welding tip and mixing assembly from the welding torch handle.

- Connect the hand cutting attachment. Tighten securely enough to prevent any possibility of gas leakage. Do *not* force it.

- Select a cutting tip appropriate for the metal thickness.

- Remove any dirt, dust, or foreign material from the tapered seat end of the cutting tip.

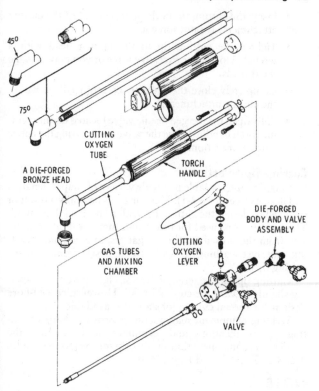

Fig. 3-2 The construction of a typical hand cutting torch.
(Courtesy Victor Equipment Co.)

- Insert the cutting tip in the mating head of the cutting attachment. Do not force it.

- Tighten the cutting tip to the mating head with a wrench. Do not use excessive force or you may damage the threads.

- Completely close the preheat oxygen needle valve on the cutting attachment.

- Fully open the oxygen needle valve located on the welding torch handle. Leave the valve open during the entire cutting operation.

Cutting Tip Selection

Cutting tips are available in a wide variety of shapes, lengths, and designs (see Table 3-1). This variety is basically a result of the need to design the cutting tip with a specific job in mind. Thus, cutting tips used for thin-gauge metal will differ in design from those used for thicker gauges, and gouging, scarfing, and rivet-washing tips will differ in design from standard cutting tips.

Cutting tips are generally of the straight-bore type, tapering slightly toward the cutting orifice. However, curved-bore designs for special cutting jobs are also available.

Welding equipment manufacturers provide charts of cutting tip specifications and recommendations to help the welder select an appropriate cutting tip for the job (see Tables 3-2, 3-3, 3-4, 3-5, and Figure 3-3).

NOTE

The drill size (oxygen cutting orifice drill size) listed in the manufacturer's chart is more important than the tip size. The latter is essentially a manufacturer's identification number and will not necessarily correlate from one manufacturer to another. However, the size of the oxygen cutting orifice is equated to a specific drill size. This can prove very helpful when an operator is confronted with the problem of matching the cutting characteristics of tips from different manufacturers.

Table 3-1 Examples of Uniweld Acetylene Cutting Tips and Their Applications

Cutting Tip	Application
	Cutting clean plate and machine cutting
	Cutting clean plate
	General purpose and machine cutting
	General purpose hand and machine cutting
	General purpose cutting

(*continued*)

Table 3-1 (continued)

Cutting Tip	Application
	Short tip for removing boiler tubes and working in confined places
	Short tip for removing boiler tubes and working in confined places
	Cutting close to bulkheads and rivet heads, and and for trimming and machine cutting
	Gouging, removing welds, rivet washing, weld preparation
	Cutting scrap with rusty or painted surfaces

(Courtesy Uniweld Products, Inc.)

Table 3-2 Cutting Tip Data Chart (Tip Size, Plate Thickness, Hose Diameter, and Gas Pressure*)

Hose Diameter	Tip Size	Metal Thickness	Fuel Gas (in psi) Acetylene	Fuel Gas (in psi) LP Gas, or Hydrogen, Natural Gas	Oxygen (in psi)
1/4" hose	68	1/8"	2	2	35
	62	3/16"–3/8"	3–5	3–5	25–32
	56	1/2"–7/8"	3–5	3–6	30–50
	53	1"–1 1/2"	3–6	4–8	35–50
	51	2"	5	8	45
3/8" hose	46	3"	6	8	40
	42	4"–6"	6–8	6–	40–55
	35	7"–8"	6–8	6–9	50–55
	30	9"–12"	8–10	7–10	55–70
	25	13"–16"	10–12	7–10	80–90

Pressures are for 50 feet of hose. If other lengths are used, pressure requirements will vary accordingly. Accurate pressures for all conditions cannot be given. Table 3-2 is intended as a guide, and the pressure ranges shown will cover the various styles of cutting tips available.

Table 3-3 Cutting Tip Data Chart (Tip size, Plate Thickness, and Cutting Speed)

Thickness of Steel Plate (inches)	Diameter of Cutting Tip Orifices (inches)	Approximate Cutting Speed Inches (per minute)
1/8	0.020–0.040	16–32
1/4	0.030–0.060	16–26
3/8	0.030–0.060	15–24
1/2	0.040–0.060	12–23
3/4	0.045–0.060	12–21
1	0.045–0.060	9–18
1 1/2	0.060–0.080	6–14
2	0.060–0.080	6–13
3	0.065–0.085	4–11
4	0.080–0.090	4–10
5	0.080–0.095	4–8
6	0.095–0.105	3–7
8	0.095–0.110	3–5
10	0.095–0.110	2–4
12	0.110–0.130	2–4

Note the following:

- A good cutting tip should demonstrate a high degree of heat resistance and long-wearing characteristics. Such tips are usually made from a copper alloy to reduce pocking or burning of the tip when slag splatters from a hot cut.

- Better cutting tips will produce sharper cuts and cleaner faces, with a minimum kerf.

- Gas delivery should be continuous and free of turbulence.

- The seating surface of the cutting tip should be precision machined for a tight, leakproof fit in the torch head.

Table 3-4 Cutting Tip Preheat Reference Heating Chart (for All Standard Victor Cutting Tips Except MTH High-Speed Cutting Models)

Tip Size	Cutting Oxygen Orifice Size″	Preheat Sizes for Various Tips											
		100	101	1-104	108	110	111	112	129	200	116	117	118
000	71 (.027″)		74 (.023)										
00	67 (.033″)		74 (.023)										71 (.027)
0	60 (.041″)	71 (.027)	74 (.023)			65 (.035)		67 (.032)		67 (.032)			
1	56 (.047″)	67 (.032)	71 (.027)		75 (.022)	60 (.040)		60 (.040)		64 (.036)			
2	53 (.060″)	60 (.040)	67 (.032)		73 (.024)	56 (.046)	64 (.036)	56 (.047)		62 (.038)	66 (.033)		63 (.037)
3	50 (.071″)		66 (.034)		66 (.033)	54 (.055)		53 (.060)	57 (.043)	60 (.040)	64 (.036)		
4	45 (.083″)		66 (.034)		63 (.037)	59 (.060)	56 (.046)	52 (.055)		56 (.046)	61 (.039)		56 (048)
5	39 (.100″)		66 (.034)		60 (.040)			52 (.055)	55 (.052)				
6	31 (.121″)		63 (.037)				53 (.060)			55 (.053)			57 (.044)
7	28 (.141″)		63 (.037)							54 (.066)			57 (.044)
8	20 (.162″)		63 (.037)										
10	13 (.188″)			55 (.052)								63 (.037)	57 (.044)
12	2 (.221″)			55 (.052)									56 (.048)

(Courtesy Victor Equipment Company)

Table 3-5 Cutting Tips Operational and Performance Data for Victor Types 100, 101, 104, 108, 110, 112, and 129 (Oxyacetylene)

Metal Thickness	Tip Size	Cutting Oxygen (PSIG)	Cutting Oxygen (SCFH)	Pre-Heat Oxygen (PSIG)	Pre-Heat Oxygen (SCFH)	Acetylene (PSIG)	Acetylene (SCFH)	Speed I.P.M.	Kerf Width
1/8"	000	20/25	20/25	3/5	3/5	3/5	3/5	28/32	0.4
1/4"	00	20/25	30/35	3/5	4/6	3/5	4/6	27/30	.05
3/8"	0	25/30	55/60	3/5	5/9	3/5	5/8	24/28	.06
1/2"	0	30/35	60/65	3/6	7/11	3/5	6/10	20/24	.06
3/4	1	30/35	80/85	4/7	9/14	3/5	8/13	17/21	.07
1"	2	35/40	140/150	4/9	11/18	3/6	10/16	15/19	.09
1-1/2"	2	40/45	150/160	4/12	13/20	3/7	12/18	13/17	.09
2"	3	40/45	210/225	5/14	15/24	4/9	14/22	12/15	.11
2-1/2"	3	45/50	225/240	5/16	18/29	4/10	16/26	10/13	.11
3"	4	40/50	270/320	6/17	20/33	5/10	18/30	9/12	.12
4"	5	45/55	390/425	7/18	24/37	5/12	22/34	8/11	.15
5"	5	50/55	425/450	7/20	29/41	5/13	26/38	7/9	.15
6"	6	45/55	500/600	10/22	33/48	7/13	30/44	6/8	.18
8"	6	45/55	500/600	10/25	37/55	7/14	34/50	5/6	.19
10"	7	45/55	700/850	15/30	44/62	10/15	40/56	4/5	.34
12"	8	45/55	900/1000	20/35	53/68	10/15	48/62	3/5	.41

(Courtesy Victor Equipment Company)

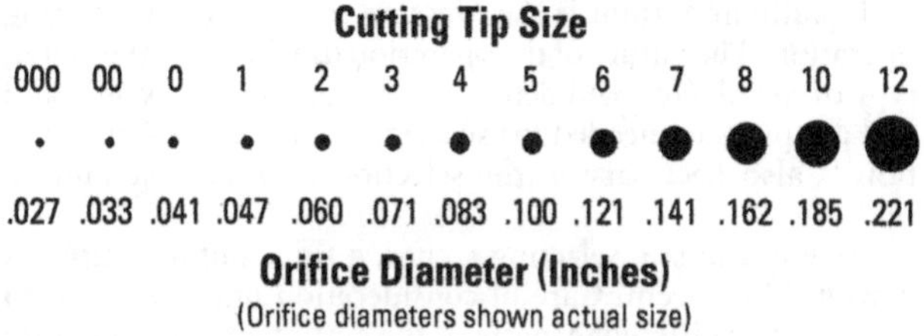

Cutting Tip Size

000	00	0	1	2	3	4	5	6	7	8	10	12
•	•	•	•	•	•	•	•	•	●	●	●	●
.027	.033	.041	.047	.060	.071	.083	.100	.121	.141	.162	.185	.221

Orifice Diameter (Inches)
(Orifice diameters shown actual size)

Fig. 3-3 Cutting Orifice Reference Chart (for all Victor Standard Cutting Tips Except MTH High-Speed Models)
(Courtesy Victor Equipment Company)

Cutting tips designed for use with natural gas or propane differ from those intended for acetylene gas. The natural gas and propane cutting tips have an outer shield into which the inner structure fits. This shield protects the flame from being blown away from the cut and increases the stability of the flame.

NOTE

Always check to make certain the equipment is rated for the selected tip size. An incorrect tip size can result in flashback. Always keep a spare tip available. Flashback arresters can provide a certain measure of protection by preventing a flashback from reaching upstream equipment.

The selection of a cutting tip is often determined by the type of fuel gas. For example, an oxyacetylene flame will require a type of cutting tip different from the one recommended for use with oxypropane or oxy-natural gas. The reason for this, of course, is that different fuel gases have different flame characteristics. Using the wrong tip can result in an unstable flame, uneconomical fuel gas consumption, and many other problems.

Equally important is the *pressure* required by the cutting operation. The nature of the operation (thickness of the metal, type of metal, etc.) will determine the amount of oxygen and fuel gas pressure needed for successful cutting. This information is also necessary in the selection of a suitable cutting tip.

As you can see, selecting a cutting tip is not as simple as it would first seem. Careful consideration must be given to a number of different aspects of the cutting operation before the final decision is made. Ultimately, experience will be the best guide for making these decisions.

Regulators

Make sure you connect an acetylene regulator to an acetylene tank and an oxygen regulator to an oxygen tank in order to accommodate the different cylinder pressures. To avoid confusion, oxygen cylinders and regulators have right-hand threads, whereas acetylene cylinders and regulators have left-hand threads.

NOTE

The female threads of a regulator must be connected to the male threads on the cylinder valve, or vice versa.

Flashback Arrestor

A flashback arrestor is a device used to prevent the burning of mixed gases inside the torch body or hoses, which can severely damage the equipment. The flashback arrestor is installed between the cutting torch and the regulator on both the oxygen and the acetylene lines (hoses). Spring-loaded valves can be used with flashback arrestors to provide increased protection against flashback. The valves are designed to detect and stop reverse gas flow.

OXYACETYLENE CUTTING PROCEDURE

Note the following:

- Hold the spark lighter slightly away from the tip and to one side when lighting the torch.

- Correct flame adjustments are very important to ensure the quality of a cut. Figure 3-4 shows the types of oxyacetylene flame adjustments recommended for manually cutting metals. Note that the oxidizing flame is not included in this group, because the oxidizing flame is used in machine cutting torches instead of hand-held ones.

- The type of metal and its thickness also determine the nature of the flame adjustment. For example, cast iron is most successfully cut with a carburizing flame; most steels are cut with a neutral flame; and thick steel castings require an oxidizing flame.

- Never allow the oxygen cutting stream to interfere with the preheating flames. Always release the cutting oxygen *after* the preheating flame has been adjusted.

- The oxygen must be at least 99.5 percent pure. A decrease in purity of only 1 percent can reduce the cutting speed by as much as 25 percent and increase gas consumption by an equal amount.

The procedure for oxyacetylene cutting may vary depending on the requirements of a specific job. However, there are certain fundamental procedural steps that can be considered common to all cutting operations. In the interest of cutting efficiency and safety, the operator should establish a set cutting procedure for himself. The procedural steps listed below are arranged in the sequence that would most likely occur in a cutting operation:

- Select a cutting tip of an appropriate size for the torch. Tip size is based on the thickness of the metal being cut.

CUTTING FLAME

Not suitable for
cutting

f. Acetylene burning in air

Excess acetylene
helps to get heat
down to the bottom
of material being
cut, this is espe-
cially suitable for
cutting cast iron.

g. Strongly Carburizing Flame — Preheat only

h. Strongly Carburizing Flame — Cutting Oxygen flowing

Standard adjust-
ment for cutting
steel.

i. Neutral Flame — Preheat only

j. Neutral Flame — Cutting Oxygen flowing

Fig. 3-4 Cutting flame adjustments. *(Courtesy Airco Welding Products.)*

- Insert the cutting tip in the torch and tighten it with the wrench recommended by the welding equipment manufacturer. Do not attempt to force the tip. If it will not tighten properly, check the heads for wear or stripping. It may need to be replaced.

- Open the fuel gas valve on the torch and purge the lines of any air.

- Adjust the fuel gas regulator to the working pressure recommended for the tip size.

- Close the fuel gas valve on the torch.

- Open the torch cutting oxygen valve.

- Adjust the oxygen regulator to the working pressure recommended for the tip size.

- Close the cutting oxygen valve. At this point, both the fuel gas and oxygen regulators have been set at their recommended working pressures. The fuel gas and cutting oxygen valves on the torch should be closed. Fuel gas and oxygen will have filled the lines down to the torch needle valves. The torch is now ready to light.

- Begin the flame adjustment by opening the fuel-gas needle valve slightly and igniting the gas with a spark lighter. Adjust the flame so that it is steady and well-defined.

- Open the oxygen needle valve (*not* the cutting oxygen valve) on the torch until a bluish-white cone is established.

- Now open the cutting oxygen valve.

- Readjust the oxygen needle valve for the correct pre-heat flame.

If the cutting stops, immediately close the cutting valve and preheat the point where the cut stopped until it is a bright red. When the cutting valve is reopened, there should be no difficulty in restarting the cut.

The heat of the preheating flame will tend to melt the edges of the cut if the torch is moved too slowly. This will produce a very ragged appearance, or at times it will fuse the metal together again.

Note the following:

- Cutting metal manually is a relatively simple procedure to learn, but the quality of the cut will depend on the skill of the operator.

- Hold the torch steadily but not tightly when hand cutting.

- Hold the torch handle with the right hand (if the operator is right-handed) and support the torch with the left hand a few inches back from the head. If the operator is left-handed, reverse the positions of the hands. In either case, the hand on the handle is the one that will operate the lever controlling the cutting oxygen.

- Do not allow the torch to waver when making the cut. Wavering results in a wider than necessary kerf (cut) and an irregular cutting speed. An irregular cutting speed will slow the speed of the cut and increase gas consumption. Both factors result in poor-quality cuts and higher operating costs.

- Watch the amount the cut curves backward away from the direction of the torch movement. The flow of the slag or oxide should be clean and unobstructed.

- Move the torch just fast enough to produce complete oxidation (combustion) of the metal along the kerf.

Cutting Steel Plates

- Adjust the preheating flame to neutral. The cutting oxygen should be shut off at this point.

- Hold the torch at a right angle (perpendicular) to the plate, with the preheating flame just touching the end of the chalk line at the edge of the steel plate (see Figure 3-5). Hold the torch steady until the spot is heated to a bright red.

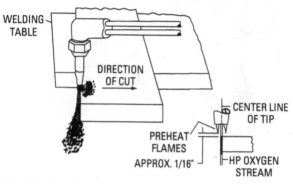

Fig. 3-5 Starting the cut. *(Courtesy Rego Company)*

- Open the valve that controls the cutting oxygen and move the torch slowly, but steadily, along the chalk line (see Figure 3-6).
- If the cut has been properly started, a shower of sparks will fall from the underside of the plate. This indicates that the cut is penetrating entirely through the plate.
- The cutting torch should be moved just fast enough that the cut continues to penetrate the plate completely (see Figure 3-7). If the torch is moved too fast, the cutting jet will fail to penetrate and the cutting will stop.
- Large, round holes can be made accurately with the aid of a radius bar or a compass. One leg of the compass is fixed at the center, and the torch nozzle makes the other leg.

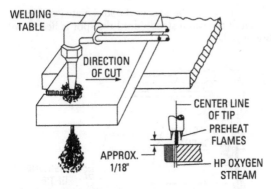

Fig. 3-6 Making the cut. *(Courtesy Rego Company)*

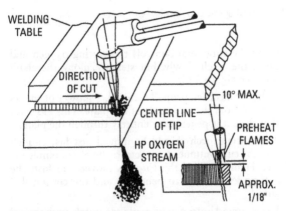

Fig. 3-7 Completing the cut. *(Courtesy Rego Company)*

Cutting Thin Sheets

The melted upper edges of thin sheets (under 1/4-inch thick) tend to fuse together during cutting. To avoid this problem, slant the torch tip and flame in the direction of the cut. The high-pressure oxygen stream will blow the melted metal out of the kerf and prevent it from fusing the kerf edges together behind the torch (see Figure 3-8).

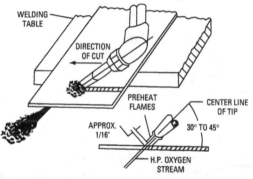

Fig. 3-8 Cutting thin sheet metal. *(Courtesy Rego Company)*

Bevel Cutting Steel

- Bevel cutting, or *beveling*, is a common operation with the cutting torch and one that should be mastered by all welders. This operation is performed by holding the torch head at an inclined angle instead of perpendicular to the work surface (vertically).

- Place a piece of 1/2-inch or thicker steel plate so that one edge extends 3 or 4 inches beyond the edge of the welding table. Incline the torch head at an angle of 45° to the top surface and cut a triangular prism off the edge. After the cut is finished, the piece of plate remaining will have one edge beveled at an angle of 45°.

- Beveling can also be done by holding the torch perpendicular to the surface. However, the preheat holes are aligned so that they are parallel to the kerf when beveling. Contrast this with the position of the preheat holes when making a straight cut, as shown in Figure 3-9.

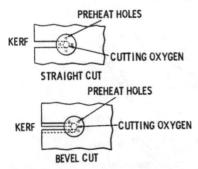

Fig. 3-9 Position of the preheat holes for a bevel cut.
(Courtesy Airco Welding Products.)

Cutting (Piercing) Holes in Steel

Sometimes cutting cannot begin at the edge of a piece. In that case, a starter hole is pierced (cut) at some point away from the edge. A straight cut is then run from the hole to the edge of the workpiece. The procedure is outlined as follows:

- Position the torch above the surface, as shown in Figure 3-10, and heat the surface until the preheating flame has produced a round, bright-red spot on the metal.
- As soon as the red spot appears (at approximately 1600°F), gradually open the high-pressure oxygen cutting valve to prevent slag from blowing back into the torch.

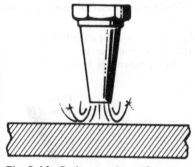

Fig. 3-10 Preheating the surface. *(Courtesy Airco Welding Products)*

- Lift the torch about 1/2 inch above the normal position used for cutting. Note: In normal cutting, the tips of the preheating flame cones are about 1/8 inch above the surface (see Figure 3-11).

- As soon as a hole begins to appear on the surface, lower the torch to the normal cutting height above the

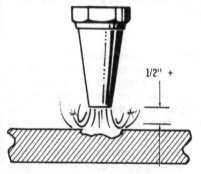

1/2" +

Fig. 3-11 Lifting the torch. *(Courtesy Airco Welding Products)*

work (1/8 inch) and hold it there until the flame pierces through the material.

- Complete a straight cut from the pierced hole to the edge of the workpiece, or from the beginning to the end of a marked surface pattern (see Figure 3-12).

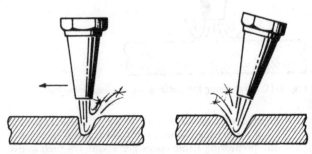

Fig. 3-12 Tilting the torch. *(Courtesy Airco Welding Products)*

NOTE

Molten slag often can be removed from the hole by slightly tilting the torch tip. It also helps to move the torch tip sideways and rotate it around the cut.

Note the following:

- The fuel-gas needle valve *must* be closed *first* when shutting off the torch. As soon as this is done, close the oxygen needle valve. The regulator pressure adjusting screws should be released last. These adjustments are for temporary shutdowns of the torch. When long periods are involved, both cylinder valves should also be closed.

- It takes longer to bring the metal surface to the cutting temperature for piercing than it does to start a cut at the edge of a workpiece.

- Do not allow slag to plug the cutting orifice on the torch. Tilt the torch slightly to blow the slag away from the tip.

- Thick metal plate should be drilled instead of pierced with a flame. Drilling is quicker and provides a smoother cutting edge.

Cutting Cast Iron

Cast iron cutting is largely limited to cutting castings apart, making repairs on cast-iron components of heavy machinery, or cutting V-grooves when preparing a cast-iron surface for welding.

Cutting cast iron requires much higher temperatures than cutting steel. The cast iron must be close to its melting point before the cut can begin. Steel, on the other hand, can be cut as soon as the metal reaches a cherry-red color. Consequently, cutting tips for cast iron are required to deliver an intense preheat. This is accomplished by using a tip with six large openings for the preheat flames.

The procedure for cutting cast iron with an oxyacetylene cutting torch should include the following steps:

- Adjust the oxygen cutting regulator so as to give the correct pressure for the thickness of the cut. This information is furnished by manufacturers of the cutting equipment.

- With the cutting valve open, adjust the heating flames to give the correct excess of acetylene, as noted in the instructions supplied by the manufacturer of the torch. Note: The amount of acetylene will vary with different makes of torches.

- The oxygen pressure is determined by the thickness of the metal to be cut. The exact oxygen pressure to be used in each case is provided in the manufacturer's information.

- Close the oxygen cutting valve.

- Begin preheating the surface with the cutting nozzle at the edge of the workpiece. As shown in Figure 3-13, the length of the excess fuel (acetylene) streamer should equal the thickness of the workpiece. In general, the hotter the cast-iron surface, the easier it will be to cut.

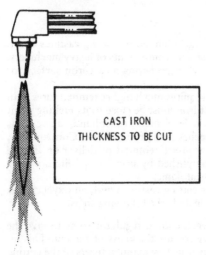

CAST IRON
THICKNESS TO BE CUT

Fig. 3-13 Length of preheating flame streamer equal to thickness of the workpiece. *(Courtesy Airco Welding Products)*

- After the preheating is accomplished, hold the torch so that the tip points backward at an angle of 75°. The inner cones of the flames should be 1/8 to 1/4 inch above the surface.
- Start the cut by swinging the torch tip in semicircles across the line of the cut (see Figure 3-14).
- Heat a semicircular area about 1/2 to 2/3 inch in diameter until the metal is actually molten. When the metal

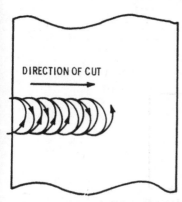

Fig. 3-14 The torch is moved in semicircular motions of 1/2 to 3/4 inch, as required to clear cut in heavy sections.
(Courtesy Airco Welding Products)

appears to boil briefly open the cutting valve to blow off the slag.

- Move the tip just off the heated edge, quickly open the cutting valve, and then move the torch along the line of cut, with the tip at a 75° angle (see Figure 3-15). Use the same swinging motion described earlier, and keep the metal hot.

- As the cut progresses, gradually raise the torch until it is at an angle of about 90° (see Figure 3-16). On heavy material, the metal will have been sufficiently preheated by the cutting process, and the cut should go forward easily enough.

- The diameter of the semicircular movement will depend on the thickness of the metal. Experience will enable the welder to reduce the diameter of the semicircular swinging motion and the width of the cut (kerf).

Fig. 3-15 Cutting tip held at a 75° angle.
(Courtesy Airco Welding Products)

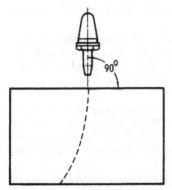

Fig. 3-16 Cutting tip held at a 90° angle.
(Courtesy Airco Welding Products)

The cutting speed should be such that the cutting oxygen jet just sweeps the edge of the cut (see Figure 3-17).

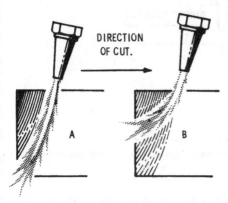

DIRECTION OF CUT.

A

B

Fig. 3-17 Cutting jet should just sweep edge of cut, as shown in (a), and not advance too deeply, as shown in (b).
(Courtesy Airco Welding Products)

- The cutting temperature can be increased by inserting a steel flux rod into the cut (as shown in Figure 3-18).

Cutting Pipe and Round Shafts
Cut or chip the surface with a chisel to provide a starting point for the torch flame. This provides a spot for the cutting torch to anchor the flame before running the cut.

Cutting Surfaces Covered with Rust, Scale, Paint, and Other Surface Contaminants
To keep particles from flying back and clogging the cutting tip, use a longer flame and hold the torch farther away from the surface of the metal. Another method is to use the torch

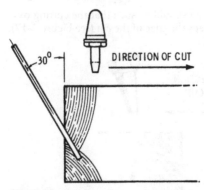

Fig. 3-18 Approximate angle at which a steel flux rod is introduced. *(Courtesy Airco Welding Products)*

flame to clean the surface along the line of the cut before beginning the cut.

Shutting Down the Equipment

CAUTION

Make sure the oxygen needle valve is closed when draining acetylene from the system. Contact between acetylene and oxygen can cause a flashback.

- Extinguish the flame by closing the acetylene needle valve on the welding torch.
- Close the preheat oxygen needle valve on the cutting attachment.
- Close the oxygen cylinder valve.
- Close the acetylene cylinder valve.
- Open the acetylene needle valve on the welding torch to drain acetylene from the system (acetylene regulator, hose, welding torch handle, and cutting attachment).

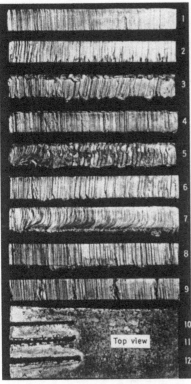

(1) Correct Procedure
Compare this cut in 1-in. plate with those below. The edge is square, the drag lines are vertical and not too pronounced.

(2) Preheat Flames Too Small
They are only about ⅛ in. long. Result: cutting speed was too slow, causing bad gouging effect at bottom.

(3) Preheat Flames Too Long
They are about ½ in. long. Result: surface has melted over, cut edge is irregular, and there is too much adhering slag.

(4) Oxygen Pressure Too Low
Result: top edge has melted over because of too slow cutting speed.

(5) Oxygen Pressure Too High
Nozzle size also too small. Result: entire control of the cut has been lost.

(6) Cutting Speed Too Slow
Result: irregularities of drag lines are emphasized.

(7) Cutting Speed Too High
Result: a pronounced rake to the drag lines and irregularities on the cut edge.

(8) Blowpipe Travel Unsteady
Result: the cut edge is wavy and irregular.

(9) Lost Cut Not Properly Restarted
Result: bad gouges where cut was restarted.

(10) Good Kerf
Compare this good kerf (viewed from the top of the plate) with those below.

(11) Too Much Preheat
Nozzle also is too close to plate. Result: bad melting of the top edges.

(12) Too Little Preheat
Flames also are too far from the plate. Result: heat spread has opened up kerf at top. Kerf is tapered and too wide.

Fig. 3-19 View of the edges of metal cut properly and improperly by the oxyacetylene process

(Courtesy Rego Company)

Table 3-6 Troubleshooting for Oxyacetylene Cutting

Problem	Possible Cause	Suggested Remedy
Flashback (flame burning in torch body, accompanied by a whistling sound).	a. Incorrect gas pressure, resulting in a low gas velocity. b. Hose leak. c. Loose hose connection. d. Torch heating flames too small. e. Cutting technique is disturbing gas flow. f. Dirt on nozzle seat.	a. Adjust gas pressure. b. Replace hose. c. Tighten, or replace if the connection is damaged. d. Adjust to larger flames. e. Change technique. f. Clean nozzle seat.
Oxygen jet stream fails to penetrate metal plate.	Torch travel speed is too slow.	Increase torch speed.
Cut lost. Bad gouges at each restart point, indicating poor restart.	Torch traveling too fast.	Turn off cutting oxygen and direct the heating flame to the end of the cut until the metal is hot enough to begin cutting again.
Top edge of the cut is melted over. Edge of the cut is irregular, with excessive amounts of slag sticking to the surface).	Preheat flames are too long.	Reduce preheat flame to appropriate length.

(continued)

Problem	Cause	Remedy
Wavy and irregular kerf wall.	Unsteady torch travel.	Adjust torch travel.
Irregular kerf, with top edge melted down.	Preheat flame is too long, resulting in excess gas being consumed by flame.	Adjust gas to reduce preheat flame.
Gouges at bottom edge of kerf, with pronounced dragline irregularities.	Low cutting speed. Preheat flames are too small.	Increase cutting speed.
Kerf too wide, resulting in wasting both gas and plate metal).	Oversize tip.	Change tip.
Piece will not drop when end of kerf is reached.	Undersize tip.	Change tip.
Rough kerf surface, with top edge melted over.	Oxygen pressure too low, causing slow cutting speed with too much lag.	Increase oxygen pressure.
Extremely rough kerf surface, with top edge melted down.	Oxygen pressure is too high and/or tip size too small.	Reduce oxygen pressure. Change torch tip if necessary.
Rough kerf with pronounced drag lines.	Torch speed is too fast.	Decrease torch speed.

- Close the acetylene needle valve when acetylene pressure drops to zero.

- Release the acetylene regulator adjustment screw by turning it counterclockwise until it rotates freely.

- Open the preheat oxygen needle valve on the cutting attachment.

- Close the preheat oxygen needle valve when oxygen pressure in the system (oxygen regulator, hose, welding torch handle, and cutting attachment) reaches zero.

- Close the oxygen needle valve on the welding torch handle.

- Release the oxygen regulator adjustment screw by turning it counterclockwise until it rotates freely.

BASIC CUTTING REQUIREMENTS

As shown in Figure 3-19, a correct cut should have a square top edge, a narrow and uniform kerf (cut width), vertical or nearly vertical drag lines (vertical ridges) on the inside surface of the kerf, and a relatively smooth kerf wall.

TROUBLESHOOTING FOR OXYACETYLENE CUTTING

There is always the possibility of trouble with anything that relies on mechanics to operate. In this case there is the torch and the hoses as well as the gages and the cylinders of gas. Table 3-6 shows some of the problems and indicates the probable cause with a suggested remedy. Check out your latest problem to see how it correlates with the items in the three columns of the Table.

4. SHIELDED METAL ARC WELDING

Shielded metal arc welding (SMAW) is a welding process in which the welding heat is produced by an electric arc formed between the work surface and a solid, consumable, covered electrode. During the welding process, the decomposition of the electrode covering forms a shield of gas and slag around the molten metal on the work surface (see Figure 4-1). The electrode also provides filler metal to the weld pool.

NOTE

Shielded metal arc welding (SMAW) is the American Welding Society's designation for this welding process. It is commonly referred to as *stick welding, arc welding,* or *stick electrode welding.* It is also sometimes referred to as manual metal arc welding (MMAC).

Some Useful SMAW Welding Tips

1. Select a proper electrode for the job in question.

2. Keep electrodes clean and dry. This is especially important for the low-hydrogen electrodes.

3. Thoroughly clean the base metal before welding. Completely remove all paint, grease, oil, moisture, slag, and any other possible weld contaminants by mechanical cleaning, chemical cleaning, or a combination of the two.

4. Use a drag technique for most applications.

5. Penetration: DCEN (least penetration), AC (medium penetration—often with more spatter than DC), DCEP (deepest penetration).

6. Use a beveled edge preparation and a multipass welding technique for materials 1/4" thick or thicker.

Shielded Metal Arc Welding Applications

Shielded metal arc welding enjoys widespread use on farms and ranches, in auto repair facilities and home and equipment

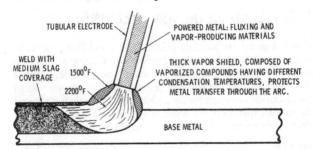

Fig. 4-1   Shielded metal arc welding process.

workshops, and in other areas where maintenance and repair work is required. It is still a widely used process in pipelining and structural steel construction.

SMAW can be used to weld carbon steel, low-alloy steel, high-strength steel, cast iron, malleable iron, bronze, nickel, stainless steels, and aluminum in all thicknesses. The SMAW process is also used for hardsurfacing.

SMAW Advantages

- Equipment is self-contained, portable, and relatively inexpensive.
- Electrode provides its own flux.
- Most metals and metal alloys can be welded with SMAW.
- Useful process for welding in confined spaces.
- Performs better on unclean surfaces than other welding processes.
- Most metal thicknesses can be welded with SMAW.
- All welding positions are possible with SMAW.
- Can be used under almost all weather conditions.
- Arc is continuously visible to the welder.
- Welder controls the arc.

SMAW Disadvantages

- Not recommended for welding metals less than 1/16" thick.

- Excessive spatter.

- Slag cleanup is required.

- Produces weld beads with rough surfaces.

- Welds are subject to porosity.

- Arc blow must be controlled.

- Frequent stops and starts are required to replace electrodes, resulting in a lower electrode deposition rate than the GMAW (MIG) and FCAW wire-feed welding processes.

- Greater possibility of weld defects as a result of frequent stops and starts.

- Some electrode waste (about 10 percent from discarded stub loss).

- Potential electric shock from open-circuit voltage.

- Ventilation must be provided when welding in confined spaces. The SMAW process produces large amounts of fumes and smoke.

SHIELDED METAL ARC WELDING EQUIPMENT

A typical shielded metal arc welding system is illustrated in Figure 4-2. Its principal components are (1) the welding machine (power source); (2) the covered electrode; (3) the electrode holder; and (4) the welding circuit cables. Additional equipment may include shielding walls and light shields; jigs, fixtures, and positioners for securing the work; and a ventilation system when working indoors.

Power Source

A constant-current (CC) power source is recommended for shielded metal. arc welding (see Table 4-1) For general

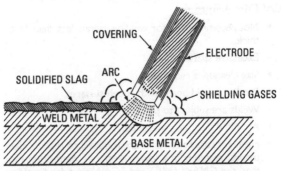

Fig. 4-2 Typical shielded metal arc welding system.

purpose welding, a welder with an AC/DC output of 225 to 300 amps works best.

CAUTION

The voltage between the terminals of the welding machine when no current is flowing in the welding circuit is called its *open-circuit voltage*. Because the open-circuit voltage can be as high as 100 volts, the welder runs the risk of severe injury or even death from electrical shock.

Power Source Grounding Requirements

- Ground the welder (power source) with a third wire in the cable connecting the circuit conductor, or by a separate wire grounded at the source current.
- Ground circuits from the welder used for purposes other than welding tools.
- Never bond either terminal of the welding generator to the frame of the welder.
- Never use pipelines containing gases or flammable liquids for a ground- return circuit.
- Never use conduits carrying electrical conductors for a ground-return circuit.

Table 4-1 Power Source Selection

Selection Factor	Comments
Location	If the welding is being done in remote areas and no electric power is available, an engine-driven (gasoline, diesel, etc.) welding machine will have to be used. If electricity is available, the type of machine selected is determined by the voltage input (most welding machines are designed to operate on either 230 or 460 volts input).
Welding current required	The welding job may call for AC, DCEN, or DCEP current. The welding machine selected should be able to deliver more amperage than the job requires.
Electrode size required	The nature of the welding job may require that a particular size of electrode be used. This will have ramifications in the amount of current used and ultimately in the type of power source selected for the job.
Amount of welding required	An automatic welding operation usually will have machine requirements different from those required by a manual or semiautomatic operation.
Metal thickness	Thicker metals require deeper penetration. This necessitates welding machines with the capacity to deliver a sufficiently high current to the electrode.

SMAW Welding Currents

Three types of welding currents are used in shielded metal arc welding: alternating current (AC), direct current electrode negative (DCEN), and direct current electrode positive

Table 4-2 AC and DC Comparisons for SMAW Welding

Current/Polarity	Comments
Alternating current (AC)	• AC arc extinguishes and restrikes 120 times a second as the current and voltage reverse direction.
	• AC current eliminates the arc blow problem encountered with direct current.
	• Produces more splatter than DC current.
	• AC equipment costs are lower than for DC.
	• Medium arc penetration.
Direct current (DC)	• DC is preferred to AC on overhead and vertical welding jobs because of its shorter and more stable arc.
	• DC current produces a more evenly distributed metal transfer than AC.
	• DC arc shorter, continuous, and more stable than the AC arc.
	• Good wetting action of the molten weld metal, and uniform weld bead size at low welding currents. For this reason it is excellent for welding thin sections.

Table 4-3 Comparing DCEN to DCEP

DC current and polarity	Comments
Direct current electrode negative (DCEN)	• DCEN produces less penetration than DCEP.
	• DCEN has a higher electrode burnoff rate than DCEP.
Direct current electrode positive (DCEP)	• DCEP produces better penetration than DCEN, but has a lower electrode burnoff rate.
	• DCEP is especially useful welding aluminum, beryllium-copper, and magnesium because of its surface cleaning action, which permits welding these metals without flux.

(DCEP). Table 4-2 and Table 4-3 compare the SMAW welding characteristics of the different currents and polarities.

NOTE

Direct current electrode negative (abbreviated DCEN or DC−) = direct current straight polarity (abbreviated DCSP). Direct current electrode positive (abbreviated DCEP or DC+) = direct current reverse polarity (abbreviated DCRP).

Electrode Holder

Electrode holders are clamping devices designed to firmly hold the electrode during the welding process. The electrode lead is commonly fastened to the end of the handle with a lug, or to a mechanical connection within the handle.

Electrode holders are manufactured in a variety of different types and sizes. The insulated pincher-type electrode holder is by far the most popular. The electrode holder size is designated by its current-carrying capacity.

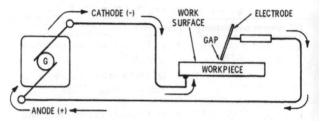

DIRECT CURRENT ELECTRODE POSITION (DCEP) WELDING CIRCUIT.

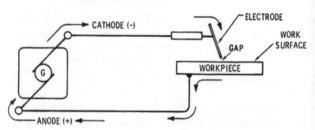

DIRECT CURRENT ELECTRODE NEGATIVE (DCEN) WELDING CIRCUIT.

Fig. 4-3 Diagram illustrating the DCEP and DCEN welding circuits. *(Courtesy James F. Lincoln Arc Welding Foundation.)*

CAUTION

The point at which the lead is connected to the electrode holder must be completely insulated, or the welder may be subjected to electric shocks.

WARNING

Sometimes an electrode holder may become uncomfortably hot during welding. This is frequently an indication of a loose connection between the lead cable and the electrode holder. It can also mean that the electrode holder is the wrong size for the work.

NOTE

The manual electrode holder must be specifically designed for arc welding and be of sufficient capacity to handle the maximum rated current required by the electrodes.

Welding Cables

Two cables are used to create the required electric circuit between the welding machine and the work. One of these cables is called the *electrode lead* and extends from the welding machine to an electrode holder. It forms one half of the electrical circuit. The other cable is called the *work lead* (or *ground lead*) and extends from the welding machine to the metal bench or the work itself. This cable forms the other half of the electrical circuit. A complete circuit is created by turning on the welding machine and bringing the electrode (in the electrode holder) in contact with the work.

Welding cables (or leads) are available in a number of different sizes. The size selected for use is dependent on the capacity of the welding machine (stated in amperes) and the maximum length of the cable. Each welding cable size represents a specific conductor (core) diameter. The conductor is that portion of the welding cable through which the electric current flows. Each diameter has a recommended *maximum* cable length. Exceeding this length without increasing the diameter of the cable results in a serious voltage drop, which will produce a poor weld. Cable diameters and their maximum recommended lengths are given in Table 4-4.

Another important factor to be considered when using welding cables is their flexibility. A flexible electrode cable is particularly important to ensure ease of movement for the welder. Flexibility is not that important for the ground cable, because it remains fairly stationary once connected to the work. The longer the cable is, the greater the required diameter. Unfortunately, as cables increase in diameter their flexibility decreases. Both of these factors must be taken into consideration when setting up an arc welding site.

Table 4-4 Cable Diameters and Their Recommended Maximum Lengths

	Cable Sizes for Lengths		
Machine Size in Amps	**Up to 50 ft**	**50–100 ft**	**100–250 ft**
75 to 100	4	2	2
200	2	1	2/0
300	0	2/0	4/0
400	2/0	3/0	4/0*
600	2/0	4/0	4/0*

Recommended longest length of a 4/0 cable for a 400-amp welder is 150 ft; for a 600-amp welder, 100 ft. For greater distances cable size should be increased, but this may be a question of cost—consider ease of handling versus moving the welding machine closer to the work.

A typical welding cable consists of:

- A conductor (core)
- A covering of rubber and reinforced fabric
- An outer jacket

The conductor consists of hundreds of fine, electrolytically pure copper strands. These are wound to form a number of wire bunches that are rope stranded to create the specific diameter size of the welding cable. Annealing the copper makes the cable soft and flexible. Welding cables are also being made with aluminum conductors, but these are not as common as the copper ones.

The conductor is covered with a layer of rubber and reinforced fabric that protects the copper from corrosion. The outer jacket consists of a durable layer of rubber or neoprene. Specifications for rubber-covered and neoprene-covered cables are listed in Table 4-5. Table 4-6 is a size selection guide for determining cable size requirements in terms of amperage usage and the distance of the work from the welding machine.

Table 4-5 Specifications for Rubber-Covered and Neoprene-Covered Welding Cable

Rubber-Covered		Neoprene-Covered		Nominal Area (Circular mils)	Stranding Nominal no. of Wires #34 AWG	Nominal Conductor Diameter (inches)	Nominal Cable OD (inches)
Cable Size	Lb per 1000 ft	Lb per 1000 ft	Ohms per 1000 ft 20°C (68°F)				
4	186	201	.225	41,740	1064	0.269	0.45
2	288	308	.163	66,360	1672	0.337	0.55
1	357	383	.129	83,690	2109	0.376	0.60
1/0	442	472	.102	105,600	2660	0.423	0.66
2/0	548	584	.084	133,100	3325	0.508	0.73
3/0	679	718	.064	167,800	4256	0.576	0.80
4/0	837	882	.051	211,600	5328	0.645	0.87

(Courtesy Airco Welding Products)

91

Table 4-6 Cable Size Selection Guide

Cable Size	Amps	50	75	100	125	150	175	200	225	250	300	350
							Distance in Feet from Welding Machine					
4	100	4	4	2	2	1	1/0	1/0	2/0	2/0	3/0	4/0
2	150	4	2	1	1/0	2/0	3/0	3/0	4/0	4/0		
	200	2	1	1/0	2/0	3/0	4/0	4/0				
1	250	2	1/0	2/0	3/0	4/0						
	300	1	2/0	3/0	4/0							
1/0	350	1/0	3/0	4/0								
	400	1/0	3/0	4/0		Based on 4-volt drop						
2/0	450	2/0	4/0									
	500	2/0	4/0									
3/0	550	3/0										
4/0	600	3/0										

(Courtesy Airco Welding Products)

NOTE

Cable size is selected based on the maximum welding current used. Sizes range from AWG No. 6 to AWG No. 4/0, with amperage ratings from 75 amperes upward. Cable length should be no longer than is required for the particular job.

SMAW ELECTRODES

SMAW electrodes are solid (or cast) wire rods covered with a thick flux coating. The thickness and composition of the flux coating determines the electrode's operating characteristics and the mechanical and chemical properties of the deposited weld. During welding the flux coating dissolves and produces a gas and slag material. The gas-and-slag shield protects the molten weld metal from contamination. When the weld deposit has cooled, the slag is removed.

Electrode Selection

It is most important to select the proper electrode for each welding job. The quality, appearance, and economy of the weld will depend upon correctly selecting the most suitable electrode (see Table 4-7). There are a number of factors that must be considered in the choice of electrode:

1. *Mechanical properties.* It is essential to know not only the kind of metal being welded (mild steel, cast iron, etc.), but also its mechanical properties. The properties of each electrode are indicated by the identification numbers of the AWS electrode classification system. Select the electrode with mechanical properties most closely matching those of the base metal. The mechanical properties should also be those that impart the highest ductility and impact resistance to the weld.

2. *Chemical properties.* The electrode should have approximately the same chemical composition as the metal being welded. The chemical properties listed for an electrode is analyses (expressed in percentages) of the different alloying elements contained in the

Table 4-7 Mild Steel Electrode Selection Guide[a]

	Electrode Class											
	E6010	E6011	E6012	E6013	E7014	E7015	E7016	E7018	E6020	E7024	E6027	E7028
Groove butt welds, flat (>1/4 in.)	4	5	8	8	9	7	7	9	10	9	10	10
Groove butt welds, all positions (>1/4 in.)	10	9	8	8	6	7	7	6	(b)	(b)	(b)	(b)
Fillet welds, flat or horizontal	2	3	7	7	9	5	5	9	10	10	9	9
Fillet welds, all positions	10	9	7	7	7	8	8	6	(b)	(b)	(b)	(b)
Current[b]	DCEP	AC DCEP	DCEN AC	DCEN DCEP AC	DCEN DCEP AC	DCEP	DCEP AC	DCEP AC	DCEN DCEP AC	DCEN DCEP AC	DCEN DCEP AC	DCEP AC
Thin material (<1/4 in.)	5	7	8	9	8	2	2	2	(b)	7	(b)	(b)
Heavy plate or highly restrained joint	8	8	6	6	8	10	10	9	8	7	8	9
High-sulfur or off-analysis steel	(b)	(b)	5	5	3	10	10	9	(b)	5	(b)	9
Deposition rate	4	4	5	5	6	4	4	6	6	10	10	9
Depth of penetration	10	9	5	6	6	7	7	7	8	4	8	8
Appearance, undercutting	6	6	6	9	9	7	10	10	9	10	10	7
Soundness	6	6	5	5	7	10	10	9	10	8	9	9
Ductility	6	7	4	5	6	10	10	10	10	5	10	9
Low-temperature impact strength	8	8	4	5	8	10	10	10	8	9	10	10
Low spatter loss	1	2	7	9	9	6	6	8	9	10	10	9
Poor fit-up	6	7	10	9	9	4	4	4	(b)	8	4	4
Welder appeal	7	6	8	9	10	6	8	8	9	10	9	9
Slag removal	9	8	6	8	8	4	4	7	9	9	9	8

*Rating is based on a comparison of same-size electrodes, with 10 as the highest value. Ratings may change with size.

†DC = either DCEP (direct current electrode positive) or DCEN (direct current electrode negative); DCEP = direct current electrode positive only; AC = alternating current. (Courtesy Airco Welding Products)

electrode's wire or core. These are nominal chemical analyses and will differ slightly for the same electrode classification among different electrode manufacturers.

3. *Welding current.* The electrode selected should be the one that most closely matches the type of power source being used. The type of welding current to be used with a particular electrode is also indicated by the AWS electrode identification numbers. As a general rule, the welder should select the maximum current (and the maximum electrode diameter) that can be used with the thickness of the metal being welded. Here, consideration should be given to whether or not the metal has been preheated. Preheated metals require less current than those that have not been preheated.

4. *Welding position.* The AWS electrode identification numbers also indicate the welding position for which the electrode is designed. Not all electrodes are designed for use in every welding position. The welder must match the electrode with the welding position being used.

5. *Thickness of the metal.* The thicker the metal, the greater the current required to produce a suitable weld. An increase in the amount of current requires a corresponding increase in electrode diameter. The welder should match, as closely as possible, the welding current being used to the electrode diameter recommended by the manufacturer.

6. *Joint design.* The design of the joint (and fitup) determines the degree of arc penetration (deep, medium, light, etc.), which is specified by the AWS electrode identification numbers. The welder should select an electrode that gives the required arc penetration.

7. *Welding passes.* The number of passes is also determined by the type of electrode selected. Multiple passes require more current than a single pass.

8. *Joint position.* The position in which joints are welded, especially on multipass work, is one of the important considerations that affect the choice of electrode. For flat and horizontal joints, the so-called *hot* electrodes should be used. Electrodes used for vertical and overhead work, of course, must produce deposits that will stay in place and not fall out of the joint while in the molten condition. Deposits of this type usually require an electrode no larger than 3/16 inch in diameter.

9. *Working conditions.* Be aware of the working conditions and select an electrode accordingly. Such factors as high temperature, low temperature, corrosive atmosphere, and impact loading are important in electrode selection.

Electrode Mechanical and Chemical Properties
The mechanical properties and chemical analyses listed for the various electrodes will vary slightly between different manufacturers for each electrode class. Always consult the electrode manufacturer's specifications when attempting to match the mechanical or chemical properties to the base metal. The mechanical properties (minimum tensile strength, yield point, etc.) will always be above the AWS minimum requirement.

Electrode Coatings
The coatings of mild and low alloy steel electrodes may have from 6 to 12 ingredients, depending on the manufacturer. Typical coatings are described in Table 4-8.

Electrode Size
The GMAW electrodes commonly range in diameter from 3/32 inch to 1/4 inch, by increments of 1/8 inch, although some manufacturers do provide diameters outside this range. Lengths are 9, 12, 14, and 18 inch, with the 14-inch length being the most common.

Table 4-8 Electrode Coatings

Coating	Comments
Alloying metals	Provides alloying content to the deposited weld metal.
Calcium fluoride	Provides shielding gas to protect the arc; adjusts slag pH level; provides both fluidity and solubility of the metal oxides.
Clays and gums	Strengthens the electrode coating; provides elasticity for extruding the electrode's plastic coating material.
Ferromanganese and ferrosilicon	Supplements the manganese and silicon content of deposited weld metal; aids in deoxidizing molten weld metal.
Iron or manganese oxide	Adjusts slag properties; adjusts slag fluidity; iron oxide in small quantities stabilizes the arc.
Iron powder	Provides additional weld metal during welding.
Metal carbonates	Provides a reducing atmosphere; adjusts slag pH.
Mineral silicates	Strengthens the electrode coating; produces slag.
Titanium dioxide	Produces a very fluid and quick-freezing, easily removable slag; provides ionization for the arc.

Electrode Current Settings

Not all manufacturers carry the same diameters for an electrode class. Furthermore, the amperage ranges for electrode diameters will vary slightly among different manufacturers. Always consult the manufacturer's electrode specification sheets for the recommended amperage range of the electrode you intend to use. It is important to remember that the current setting will be affected not only by the thickness of the material being welded but also by the welding position. For example, the amperage ranges for different diameters of an electrode in the vertical and overhead positions will be slightly lower than amperage ranges in the flat position. The amperage current ranges and electrode diameters listed in Table 4-9 are offered as a basis for comparison.

Electrode Classification

All mild steel and low alloy electrodes are classified with a four- or five-digit number prefixed by the letter E. These so-called *stick electrodes* are no longer color-coded for identification. The AWS classification is stamped on the electrode coating surface and is described in Tables 4-10 through 4-14.

Electrode Operating Characteristics

Electrodes can be divided by their operating characteristics into three categories: fast-freeze electrodes, fill-freeze electrodes, and fast-fill electrodes (Tables 4-15, 4-16 and 4-17). There will be cases where a joint may require an electrode with a combination of these categories.

Low-Hydrogen Electrodes

The presence of hydrogen during the welding seriously affects the quality of the weld. Electrodes with low-hydrogen coatings were developed to eliminate this problem. These coatings have a very low moisture content, a factor which prevents the introduction of hydrogen into the weld during the welding process. The E6015, E6016, and E6018 electrodes are the original low-hydrogen electrodes.

Iron Powder Electrodes

Faster deposition rates can be obtained by using electrodes with significant amounts of iron powder in their coating. The percentage of iron in the coating is set at either 30 percent (lime coating) or 50 percent (titania coating). The 50 percent iron powder content allows more current to be used than the 30 percent content. It produces a much faster weld, but is limited to the flat (downhand) welding position. Because the coating is thicker than other types these electrodes require higher currents.

Table 4-9 Typical Amperage Current Ranges for Various Electrode Diameters

AWS Classification	Electrode Current/Polarity	Electrode Diameters (in inches) and Current Ranges (in Amperes)							
		1/16	5/64	3/32	1/8	5/32	3/16	7/32	1/4
E6010	DCEP	—	—	60–85	80–120	110–169	150–200	200–275	220–325
E6011	AC	—	—	65–90	80–120	130–170	170–210	150–260	190–300
E6012	DCEP	—	—	40–75	70–110	80–145	110–180	135–235	170–270
E6013	DCEN	—	—	—	80–135	110–180	155–250	225–295	245–325
	AC	25–50	35–60	—	90–150	120–200	170–270	250–325	275–360
	DCEP	—	—	50–100	80–130	140–180	180–230	—	—
E7014	AC	—	—	70–95	100–135	145–180	190–235	—	—
	DCEN	—	—	50–100	90–140	150–210	200–240	260–340	280–425
	DCEP	—	—	75–95	100–145	135–200	185–235	235–305	260–380
E7015	DCEP	—	—	65–110	100–150	140–200	180–255	240–320	300–390
E7016	DCEP	—	—	65–110	100–150	140–200	180–255	240–320	300–390
E7018	DCEP	—	—	55–85	90–140	130–185	190–250	210–330	290–430
	AC	—	—	80–120	110–170	135–225	200–300	260–380	325–530
E6020	DCEN	—	—	—	120–160	130–190	175–250	225–310	275–375
	DCEP	—	—	—	—	—	—	—	—
	AC	—	—	—	—	—	—	—	—

(continued)

Table 4-9 (continued)

AWS Classification	Electrode Current/Polarity	Electrode Diameters (in inches) and Current Ranges (in Amperes)							
		1/16	5/64	3/32	1/8	5/32	3/16	7/32	1/4
E7024	DCEN	—	—	—	115–175	180–240	240–315	300–380	350–450
	DCEP	—	—	—	12–160	160–215	215–285	270–340	315–405
	AC	—	—	—	—	—	—	—	—
E6027	DCEN	—	—	—	—	190–240	250–300	300–380	350–450
	DCEP	—	—	—	—	175–215	230–270	270–340	315–405
	AC	—	—	—	—	—	—	—	—
E7028	DCEP	—	—	—	—	180–270	240–330	275–410	360–520
	AC	—	—	—	—	170–240	210–300	60–380	—

Table 4-10 Explanation of AWS Classification Numbers for Mild Steel and Low Alloy Electrodes

AWS Classification	Explanation
E	Letter prefix designating an electrode.
6	Two numbers indicating the required minimum
0	tensile strength in 1000-psi units. **Note:** Some electrodes have five numbers after the letter prefix—for example, an E12018 electrode. In this case, the first three digits indicate the minimum tensile strength—120,000 psi for this electrode
1	Third digit indicating the welding position.
1	Fourth digit indicating the type of electrode coating and welding current (see Table 4-11).

NOTE

The iron content can be determined within certain ranges by the last digit of the electrode identification number. EXX4 and EXX8 both indicate an iron powder content of 30 to 50 percent. Electrodes having identification numbers of EXXO and EXX2 contain 0 to 10 percent iron powder, and an EXX7 number indicates a 50 percent iron-powder content.

SMAW ALLOY STEEL ELECTRODES

The metal core of these electrodes is made of alloy steel instead of low carbon steel. The electrode coating is similar to the low-hydrogen type. Some may also contain iron powder. They are designed for welding high strength alloy steels and can

Table 4-11 Third Digit: Welding Position

Third Digit	Welding Position
EXX1X	All positions: flat, horizontal, vertical, overhead.
EXX2X	Flat and horizontal positions only.
EXX3X	Flat, horizontal, vertical down, overhead positions.

Table 4-12 Fourth Digit: Type of Electrode Coating and Current

Fourth Digit	Type of Coating	Welding Current
EXXX0	Cellulose sodium	DCEP
EXXX1	Cellulose potassium	AC or DCEP or DCEN
EXXX2	Titania sodium	AC or DCEN
EXXX3	Titania potassium	AC or DCEP
EXXX4	Iron powder titania	AC or DCEN or DCEP
EXXX5	Low-hydrogen sodium	DCEP
EXXX6	Low-hydrogen potassium	AC or DCEP
EXXX7	Iron powder iron oxide	AC or DCEP or DCEN
EXXX8	Iron powder low hydrogen.	AC or DCEP

AC = alternating current; DCEP = direct current electrode positive; DCEN = direct current electrode negative.

deposit welds with a tensile strength in excess of 100,000 psi. Common applications include the welding of high temperature, high pressure piping, carbon moly piping used in high pressure, high temperature steam service, and plates or castings with a molybdenum content of approximately 0.50 percent.

Electrodes in the E70XX series (E7010, E7011, E7013, E7015, E7016, E7020, E7025, E7026 and E7030) are commonly referred to as *low-alloy steel electrodes*. The E8010, -11, -13; E9010, -11, -13; and E10010, -11, -13 electrodes also belong to this group.

SMAW STAINLESS STEEL ELECTRODES

Stainless-steel electrodes are available with either lime or titania coatings. The first is used only with DC electrode positive (DCEP), the second can be used with both AC and DC electrode positive (DCEP) current.

The lime-coated electrodes produce flat or slightly convex fillet welds. The slag covers the entire weld, spatter is at a

Table 4-13 Two-Digit Suffix Indicating Chemical Composition of Weld Deposit

Suffix	MN (%)	Ni (%)	Cr (%)	MO (%)	V (%)
EXXXX-A1				.50	
EXXXX-B1			.50	.50	
EXXXX-B2			1.25	.50	
EXXXX-B3			2.25	1	
EXXXX-C1		2.50			
EXXXX-C2		3.25			
EXXXX-C3		1	.15	.35	
EXXXX-D1 & EXXXX-D2	1.25–2.00				
EXXXX-G	.50	.50	.30 (min.)	.20 (min.)	.10 (min.)

Table 4-14 Optional Supplemental Hydrogen Designators (Maximum Diffusible Hydrogen Level Obtained with the Electrode)*

Supplemental Designator	Comments
EXXXX-XX H8	Electrodes or electrode–flux combinations capable of depositing weld metal with a maximum diffusible hydrogen content of 8 mL/100 g (H8).
EXXXX-XX H4	Electrodes or electrode–flux combinations capable of depositing weld metal with a maximum diffusible hydrogen content of 4 mL/100 g (H4).
EXXXX-XX H3	Electrodes or electrode–flux combinations capable of depositing weld metal with a maximum diffusible hydrogen content of 3 mL/100 g (H3). Uncontrolled hydrogen level.
EXXXX-XX H2	Electrodes or electrode–flux combinations capable of depositing weld metal with a maximum diffusible hydrogen content of 2 mL/100 g (H2). Low hydrogen level.
EXXXX-XX H1	Electrodes or electrode–flux combinations capable of depositing weld metal with a maximum diffusible hydrogen content of 1 mL/100 g (H1). Extra-low hydrogen level.
EXXXX-XX H4R	*R* letter indicates the ability of the electrode to meet specific low-moisture pickup limits under controlled humidification tests.

*Table created by author.

Table 4-15 Fast-Freeze Electrodes

Operating Characteristics	Comments
• Weld metal rapidly solidifies in all positions	General, all-purpose electrode. Recommended for welding in the vertical and overhead positions.
• Slow deposition rate.	
• Deep penetration with maximum admixture	
• Easily removed light slag cover	
• Flat beads with distinct ripples	

Table 4-16 Fill-Freeze Electrodes[1]

Operating Characteristics	Comments
• Flat or slightly convex weld bead with distinct ripples	The weld metal freezes rapidly even though the slag remains relatively fluid.
• Medium deposition rate	
• Little spatter	

[1] Also sometimes called "fast-follow electrodes."

Table 4-17 Fast-Fill Electrodes

Operating Characteristics	Comments
• Fastest deposition rate.	Used to fill joints that require high deposition rates in the shortest time.
• Shallow penetration.	
• Minimum spatter with easily removed slag.	
• Smooth, ripple-free weld bead.	
• Flat or slightly convex bead appearance.	

minimum, and the impurities are fluxed from the weld metal. The titania-coated electrodes produce slightly concave type welds with a smoother and more stable arc than that found with the lime-coated type.

The numbering identification system for stainless steel electrodes differs from the one used for mild steel and low-alloy steel electrodes. The prefixed *E* indicates an arc welding electrode. The three-digit number following the prefix letter indicates the type of stainless steel (e.g., 308, 410, 502, etc.). Two more digits follow and are separated from the first three by a hyphen. These last two digits indicate the coating, current polarity, and welding position of each electrode.

- -15 lime coating, DCEP in all welding positions.
- -16 titania coating, AC or DCEP in all welding positions.
- -25 titania coating, DCEP in horizontal and flat positions.
- -26 titania coating, AC or DCEP in horizontal and flat positions

SMAW ALUMINUM ELECTRODES
Aluminum electrodes commonly used for welding aluminum are ER1100, ER4043, ER5554, ER5356, ER5556, and ER5183. The ER1100 and the ER4043 are the most widely used aluminum electrodes. ER4043 enjoys a reputation as being a very useful general-purpose electrode. Some commonly used SMAW aluminum electrodes are listed in Appendix C.

SMAW SPECIALIZED ELECTRODES
Specialized electrodes have been developed for a wide variety of different metals and metal alloys. These electrodes are identified by the prefix *E* (for electrode) plus letters indicating their principal alloying element and then followed by letters listing the remaining content. Specialized electrodes are grouped according to the principal alloying element (Table 4-18). Some commonly used SMAW specialized electrodes are listed in Appendix D.

Table 4-18 Examples of Common Specialized Electrode Designations

Group	AWS Electrode Classification	Electrode Name	Composition (Typical Analysis)
Copper and Copper Alloy Electrodes	Ecu	Deoxidized Copper	Copper (Cu)
	EcuAl	Aluminum Bronze	Copper (Cu), Aluminum (Al)
	EcuSi	Silicon Bronze	Copper (Cu), Silicon (Si)
	ECuSn	Phosphor Bronze	Copper (Cu), Tin (Sn)
	ECuNi	Copper Nickel	Copper (Cu), Nickel (Ni)
	ECuNiAl	Nickel Aluminum Bronze	Copper (Cu), Nickel (Ni), Aluminum (Al)
	ECuMgNiAl	Magnesium Nickel Aluminum Bronze	Copper (Cu), Magnesium (Mg), Nickel (Ni), Aluminum (Al)
Nickel and Nickel Alloy Electrodes	ENiCu	Nickel Copper	Nickel (Ni), Copper (Cu)
	ENiCrFe	Nickel Chromium Iron	Nickel (Ni), Chromium (Cr), Iron (Fe)
	ENiCrMo	Nickel Chromium Molybdenum	Nickel (Ni), Chromium (Cr), Molybdenum (Mo)

NOTE

An *ER* prefix used with specialized electrodes indicates an electrode wire used in the GTAW (TIG) or GMAW (MIG) processes.

SMAW TROUBLESHOOTING

Table 4-19 MAW Power Source Troubleshooting

Problem	Possible Cause	Suggested Remedy
Welding machine runs for a short time and then stops.	a. Wrong relay heaters.	a. Replace relay heaters.
	b. Welder is overloaded.	b. Considerable overload can be carried only for a short time.
	c. Duty cycle is too high.	c. Do not operate continually at overload currents.
	d. Leads are too long or too narrow in cross-section.	d. Should be large enough to carry welding current without excessive voltage drop.
	e. Power circuit is single phased.	e. Check for one dead fuse or line.
	f. Ambient temperature is too high.	f. Operate at reduced loads where temperature exceeds 100°F.
	g. Ventilation is blocked.	g. Check air inlet and exhaust openings.
Loud arc with excessive spatter.	a. Current setting too high.	a. Check setting and output with ammeter.
	b. Wrong polarity.	b. Check polarity; try reversing it, or use an electrode of opposite polarity.

(continued)

Table 4-19 (continued)

Problem	Possible Cause	Suggested Remedy
Sluggish arc.	a. Current is too low.	a. Check the output and current recommended for the electrode being used.
	b. Poor connection.	b. Check all electrode holder, cable, and ground-cable connections.
	c. Cable is too long or too small.	c. Check cable voltage drop and change cable.
Generator control fails to vary current.	Any part of field circuit may be short-circuited or open-circuited.	Find faulty contact and repair it.
Welding machine starts but fails to deliver required current.	a. Worn, damaged, or missing brushes.	a. Check that all brushes bear on commutator with sufficient tension.
	b. Loose brush connections.	b. Tighten them.
	c. Open field circuit.	c. Check connection to rheostat, resistor, and auxiliary brush studs.
	d. Open series field and armature circuit.	d. Check it with test lamp or bell ringer.
	e. Wrong driving speed.	e. Check nameplate against speed of motor or drive.

(continued)

Table 4-19 (continued)

Problem	Possible Cause	Suggested Remedy
	f. Dirt on grounding field coils.	f. Clean and reinsulate.
	g. Welding terminal shorted.	g. electrode holder or cable grounded.
Welding machine producing current, but current falls off during welding.	a. Loose electrode or ground connection.	a. Clean and tighten all connections.
	b. Poor ground.	b. Check ground-return circuit.
	c. Brushes worn off.	c. Replace with recommended grade. Sand to fit. Blow out carbon dust.
	d. Weak brush spring.	d. Replace or readjust brush springs.
	e. Weak brush spring pressure.	e. Replace or readjust brush springs.
	f. Brush not properly fitted.	f. Sand brushes to fit.
	g. Brushes installed backwards.	g. Reverse.
	h. Wrong brushes used.	h. See renewal part recommendations.
	i. Brush pigtails damaged.	i. Replace brushes.
	j. Rough or dirty commutator.	j. Turn down or clan commutator.
	k. Motor connection single-phased.	k. Check all connections.

(continued)

Table 4-19 (continued)

Problem	Possible Cause	Suggested Remedy
Starter not operating— machine will not start.	a. Power circuit is dead.	a. Check voltage.
	b. Broken power lead.	b. Repair.
	c. Wrong supply voltage.	c. Check nameplate against supply.
	d. Open power switches.	d. Close.
	e. Blown fuses.	e. Replace.
	f. Overload relay tripped.	f. Let set cool. Remove cause of overloading.
	g. Open circuit to starter button.	g. Repair.
	h. Defective operating coil.	h. Replace it.
	i. Mechanical obstruction in contactor.	i. Remove it.
Starter operating— machine will not start.	a. Wrong motor connections.	a. Check connection diagram.
	b. Wrong supply voltage.	b. Check nameplate against supply.
	c. Rotor stuck.	c. Try turning it by hand.
	d. Power circuit is single-phased.	d. Replace fuse; repair open line.
	e. Starter is single-phased.	e. Check contact of starter tips.
	f. Poor motor connection.	f. Tighten it.
	g. Open circuit in windings.	g. Repair it.

(continued)

Table 4-19 (continued)

Problem	Possible Cause	Suggested Remedy
Starter operates and blows fuse.	a. Fuse is too small.	a. Should be two to three times rated motor circuit.
	b. Short circuit in motor connections.	b. Check starter and motor leads for insulation from ground and from each other.

Table 4-20 SMAW Weld Troubleshooting

Problem	Possible Cause	Suggested Remedy
Porous weld	a. Arc too short	a. Use and hold longer arc.
	b. Wrong electrode size or type.	b. Select correct electrode for application.
	c. Gases trapped by insufficient puddling time.	c. Allow sufficient puddling time for gases to escape.
	d. Faulty electrode.	d. Replace electrode.
Poor penetration	a. Speed too fast.	a. Reduce welding speed. Use just enough current to obtain desired penetration.
	b. Electrode too large.	b. Match electrode to groove size.
	c. Welding speed too low.	c. Increase welding speed and use enough current for to provide sufficient arc penetration.
	d. Incorrect gap size at bottom of weld.	d. Adjust gap size.

(continued)

Table 4-20 *(continued)*

Problem	Possible Cause	Suggested Remedy
Undercutting	a. Current too high (often accompanied by excessive spatter and arc blow).	a. Lower the welding current until spatter and arc blow are corrected and undercutting ceases.
	b. Incorrect electrode movement or position.	b. Use a uniform weave in butt welding. Avoid excessive weaving. Hold electrode at safe distance from vertical plane in making horizontal fillet weld.
	c. Incorrect electrode diameter.	c. Use correct electrode diameter for the job. Undercutting usually occurs when the electrode diameter is too large.
	d. Arc blow.	d. See *Poor fusion*.
Cracked weld	a. Incorrect electrode.	a. Chose proper electrode for part size.
	b. Weld and part sizes do not match.	b. Adjust weld size to part size.
	c. Incorrect joint gap size for weld.	c. Establish correct and uniform gap size.
	d. Amperage too high.	d. Use amperage as low as possible for sound weld.

(continued)

Table 4-20 (continued)

Problem	Possible Cause	Suggested Remedy
Poor weld appearance	a. Faulty or incorrect electrode.	a. Replace electrode.
	b. Improper use of electrode.	b. Use a proper welding technique. Use a uniform weave.
	c. Wrong arc voltage and current.	c. Avoid overly high voltage.
	d. Overheating.	d. Avoid overheating. Maintain heat within proper heat range.
Poor fusion	a. Arc blow (wavering of the arc from its intended path).	a. Reduce the current; or use back stepping method on long welds; or touch surface with electrode coating and direct the electrode tip in the direction opposite to that of the arc blow; or use AC and appropriate electrode.
	b. Current improperly adjusted.	b. Select proper current and voltage.
	c. Improper electrode size.	c. Select electrode to match joint.

(continued)

Table 4-20 (continued)

Problem	Possible Cause	Suggested Remedy
Brittle weld	a. Faulty or incorrect electrode.	a. Replace with correct electrode.
	b. Faulty preheating.	b. Preheat at 300°F if welding on medium carbon steel or certain alloy steels.
	c. Stress causing brittleness.	c. Relieve stress after welding.
Excessive spatter	a. Arc blow.	a. See *poor fusion*.
	b. Arc too long.	b. Adjust for shorter arc length
	c. Current too high.	c. Adjust current properly.
	d. Faulty or incorrect electrode.	d. Replace electrode.

5. ARC WELDING AND CUTTING

From the time electricity was available commercially, it was known that an electric arc between two electrodes was a concentrated heat source that reached a temperature of 3871°C (7000°F). In 1881, various persons attempted to use an arc between a carbon electrode and metal workpieces. The heat source for fusion welding using a basic electrical circuit is shown in Figure 5-1.

In gas-flame welding it was necessary to add filler metal from a metallic wire. Early arc welding then used a bare metal wire as the electrode. It melted in the arc and thus automatically supplied the needed filler metal. Results were very uncertain, however The arc was unstable, and a great amount of skill was required to maintain it. Contamination and oxidation of the weld metal resulted from exposure to the atmosphere at such high temperatures. At that time, there was little or no understanding of the metallurgical effects and requirements for welding.

The great potential of the arc welding process was recognized, however, especially as a result of experiences provided by World War I. There was very little use made of welding until after the war. Then, shielded metal electrodes were developed in the 1920s. They provided a stable arc and shielding from the atmosphere, but some fluxing action for the molten pool of metal was needed. The major problems related to arc welding were solved as a number of variations were introduced and developed for different materials and processes, especially in the manufacturing industries. See Figure 5-1.

Arc welding is done with metal electrodes. In one type, the electrode is consumed as it supplies the needed filler metal to fill the voids in the joint, thereby speeding the welding process. In this case the electrode has a melting point below the temperature of the arc. Small droplets are melted from the end of the electrode and pass to the parent metal. The size of these droplets varies greatly, and the mechanism of transfer

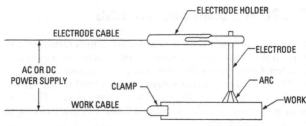

Fig. 5-1 A basic welding circuit.

varies with different types of electrodes and processes. As the electrode melts the arc length changes, and the resistance of the arc path varies with the change in the arc length and the presence—or absence—of metal particles in the path. This requires that the electrode be moved toward or away from the work to maintain the arc and satisfactory welding conditions.

No ordinary manual arc welding is done today with bare electrodes; shielded (covered) electrodes are used instead. However, a large amount of automatic and semiautomatic arc welding is done in which the electrode is a continuous, bare metal wire. This method requires a separate shielding and arc-stabilizing medium. Automatic feed-control devices maintain the correct arc length.

There are other types of metal arc welding that use an electrode made of tungsten. The tungsten is not consumed by the arc, except by relatively slow vaporization. In some applications, a separate filler wire must be used to supply the needed metal.

SHIELDED METAL ARC WELDING

Shielded metal arc welding (SMAW) uses electrodes made of metal wire, usually from 1.59 to 9.53 mm (1/16 to $^3/_8$ inch) in diameter. The wire's extruded coating has chemical components that add desirable characteristics, including all or a number of the following:

- Provide a protective atmosphere
- Stabilize the arc
- Act as a flux to remove impurities from the molten metal
- Provide a protective slag to accumulate impurities, prevent oxidation, and slow down the cooling of the weld metal
- Reduce weld-metal spatter and increase the efficiency of deposition
- Add alloying elements
- Affect arc penetration
- Influence the shape of the weld bead
- Add additional filler metal

All arc-welding processes use essentially the same basic circuit. Alternating current is used as much as direct current today. When direct current is used, if the work is made positive (+) as the anode of the circuit, and the electrode is made negative (−), the condition is called *straight polarity.* When the work is negative and the electrode is positive, the polarity is *reversed.* Using bare electrodes liberates greater heat at the anode, but certain shielded electrodes change the heat conditions and are used with reverse polarity.

Coated electrodes are classified on the basis of the tensile strength of the deposited weld metal, the welding position in which they may be used, the type of current and polarity (if direct current), and the type of covering. A four- or five-digit system of designation is used, as indicated in Figure 5-2. As an example, type E7016 is a low-alloy steel electrode that will provide a deposit having a minimum tensile strength of 70,000 psi in the non-stress-relieved condition. This eletrode can be used in all positions with either alternating current or reverse-polarity direct current, and it has a low-hydrogen type of coating (see Figure 5-3).

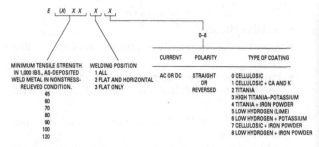

Fig. 5-2 Arc-welding electrode designation system.

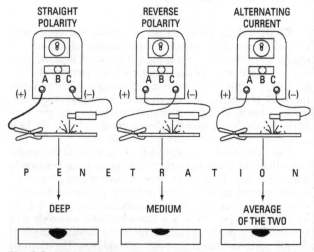

Fig. 5-3 Welding polarity is one of the factors that control the penetration of the metal deposit.

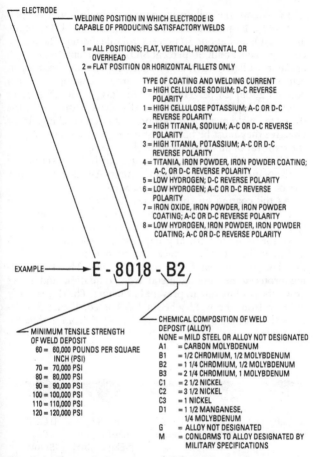

ELECTRODE

WELDING POSITION IN WHICH ELECTRODE IS
CAPABLE OF PRODUCING SATISFACTORY WELDS

1 = ALL POSITIONS; FLAT, VERTICAL, HORIZONTAL, OR
    OVERHEAD
2 = FLAT POSITION OR HORIZONTAL FILLETS ONLY

TYPE OF COATING AND WELDING CURRENT
0 = HIGH CELLULOSE SODIUM; D-C REVERSE
    POLARITY
1 = HIGH CELLULOSE POTASSIUM; A-C OR D-C
    REVERSE POLARITY
2 = HIGH TITANIA, SODIUM; A-C OR D-C REVERSE
    POLARITY
3 = HIGH TITANIA, POTASSIUM; A-C OR D-C
    REVERSE POLARITY
4 = TITANIA, IRON POWDER, IRON POWDER COATING;
    A-C, OR D-C REVERSE POLARITY
5 = LOW HYDROGEN; D-C REVERSE POLARITY
6 = LOW HYDROGEN; A-C OR D-C REVERSE
    POLARITY
7 = IRON OXIDE, IRON POWDER, IRON POWDER
    COATING; A-C OR D-C REVERSE POLARITY
8 = LOW HYDROGEN, IRON POWDER, IRON POWDER
    COATING; A-C OR D-C REVERSE POLARITY

EXAMPLE — E - 8018 - B2

MINIMUM TENSILE STRENGTH
OF WELD DEPOSIT
60 = 60,000 POUNDS PER SQUARE
     INCH (PSI)
70 = 70,000 PSI
80 = 80,000 PSI
90 = 90,000 PSI
100 = 100,000 PSI
110 = 110,000 PSI
120 = 120,000 PSI

CHEMICAL COMPOSITION OF WELD
DEPOSIT (ALLOY)
NONE = MILD STEEL OR ALLOY NOT DESIGNATED
A1 = CARBON MOLYBDENUM
B1 = 1/2 CHROMIUM, 1/2 MOLYBDENUM
B2 = 1 1/4 CHROMIUM, 1/2 MOLYBDENUM
B3 = 2 1/4 CHROMIUM, 1 MOLYBDENUM
C1 = 2 1/2 NICKEL
C2 = 3 1/2 NICKEL
C3 = 1 NICKEL
D1 = 1 1/2 MANGANESE,
      1/4 MOLYBDENUM
G   = ALLOY NOT DESIGNATED
M   = CONLORMS TO ALLOY DESIGNATED BY
      MILITARY SPECIFICATIONS

Fig. 5-4 Electrode designation as favored by the American Welding Society.

In general, the cellulosic coatings contain about 50 percent SiO_2; 10 percent Ti_2O; small amounts of FeO, MgO, and Na_2O; and about 30 percent volatile matter. The titania coatings have about 30 percent SiO_2, 50 percent Ti_2O; small amounts of FeO, MgO, Na_2O, and Al_2O_3; and about 5 percent volatile material. The low-hydrogen coatings typically contain about 28 percent Ti_2O plus ZrQ_2, and 25 percent CaO plus MgO. They eliminate dissolved hydrogen in the deposited weld metal and thus prevent microcracking. To be effective they must be baked just prior to use, which ensures the removal of all moisture from the coating.

All electrodes are marked with standard colors established by the National Electrical Manufacturers Association, so they can be readily identified as to type. See Figure 5-4.

As the coating on the electrode melts and vaporizes, it forms a protective atmosphere that stabilizes the arc and protects the molten and hot metal from contamination. Fluxing constituents unite with any impurities in the molten metal and float them to the surface, where they are entrapped in the slag coating that forms over the weld. This slag coating protects the cooling metal from oxidation and slows down the cooling rate to prevent hardening. The slag is easily chipped from the weld when it has cooled. Figure 5-5 depicts the way in which metal is deposited from a shielded electrode.

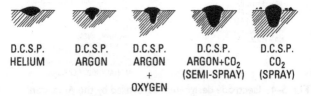

D.C.S.P.
HELIUM

D.C.S.P.
ARGON

D.C.S.P.
ARGON
+
OXYGEN

D.C.S.P.
ARGON+CO2
(SEMI-SPRAY)

D.C.S.P.
CO2
(SPRAY)

Fig. 5-5 Weld penetration changes for different shielding gases, polarity, and amount of current applied.

Electrodes having iron powder in the coating are used extensively, particularly in production-type welding. This reduces welding costs by greatly increasing the amount of metal that can be deposited with a given size of electrode wire and amount of current.

The insulating coating on one type of electrode melts slowly enough to protrude slightly beyond the melting filler wire, and if dragged along the work it will maintain the proper arc length. These are called *contact* or *drag electrodes*.

Although a very large amount of welding still is done with ordinary shielded electrodes, in recent years there has been a great increase in the use of other methods of shielding, largely because they permit the use of continuous electrodes and automatic electrode-feeding devices.

GAS TUNGSTEN ARC WELDING

Gas tungsten arc welding (GTAW) was one of the first major developments away from the use of ordinary shielded electrodes. It was developed in the 1940s for metals that were difficult to weld. Aluminum, magnesium, chrome, and molybdenum steels were of concern because they were used in aircraft and other war materials. Originally developed for welding magnesium, GTAW uses a tungsten electrode held in a special holder. The diameter of the tungsten electrode is changed to fit the job. An inert gas passes through the holder with sufficient flow to form an inert shield around the arc and the molten pool of metal. This shield protects the molten pool from the atmosphere. The arrangement is shown in Figure 5-6.

GTAW will weld most metals and alloys, including:

- alloy steels
- carbon steels
- stainless steels

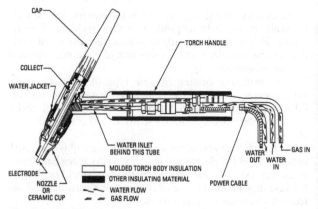

Fig. 5-6   Welding torch used in inert gas (GTAW) welding with a nonconsumable metal electrode.

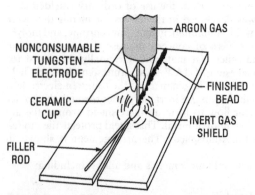

Fig. 5-6 (a)   The gas tungsten arc produces its welding heat by holding an arc between the tungsten rod tip and the base metal. Note that the high temperature area is protected by an inert gas.

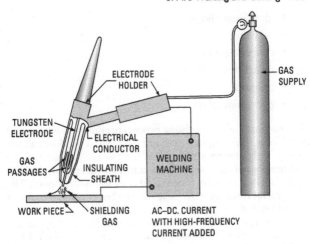

Fig. 5-6(b) Basic equipment needed for gas metal arc welding.

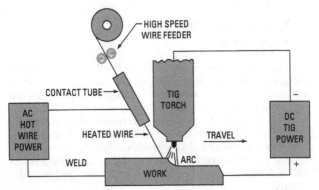

Fig. 5-6(c) Schematic diagram of the hot-filler-wire (GTAW) welding process.

- aluminum alloys
- beryllium alloys
- copper-based alloys
- nickel-based alloys
- titanium
- zirconium alloys

The gas tungsten arc welding process produces a high-quality weld joint. The tungsten does not melt, but it produces an arc that is a clean source of heat. When the arc has produced a molten pool of base metal that is under the gas shield, the filler rod is melted into the forward edge of the molten pool. The filler metal fuses with the base metal, then cools and makes a continuous weld.

The GTAW process can be used manually or set up for automatic welding. All welding processes can be done by GTAW, including:

- continuous beads
- skip welds
- intermittent welds
- spot welds

Argon or helium, or a mixture of them, is used as the inert gas shielding medium. Because the tungsten electrode is not consumed at arc temperatures in these inert gases, the arc length remains constant so that the arc is stable and easy to maintain. The tungsten electrodes often are treated with thorium or zirconium to provide better current-carrying and electron-emission characteristics. A high-frequency, high-voltage current usually is superimposed on the regular AC or DC welding current to make it easier to start and maintain the arc.

When a filler wire is needed, a fine, continuous filler wire is heated by passing an electric current through it. The wire melts as it feeds into the weld puddle just behind the arc, as a result of the I2R effect. I2R drop means electrically it is the current squared time the resistance that produces the heat for welding. See Ohm's Law for more detailed info. The deposition rate is several times what can be achieved with a cold wire, and it can be increased further by oscillating the filler wire from side to side when making a wide weld. This hot-wire process cannot be used in welding copper or aluminum, however, because of their inherently low resistances.

As mentioned earlier, gas tungsten arc welding produces very clean welds. No cleaning or slag removal is required because no flux is used. Skilled operators often produce welds that can scarcely be seen. However, the surfaces to be welded must be clean and free of oil, grease, paint, or rust, because the inert gas does not provide any cleaning or fluxing action.

Gas Tungsten Arc Spot Welding

A variation of gas tungsten arc welding is employed for making spot welds between two pieces of metal without the necessity of having access to both sides of the joint. The basic procedure is shown in Figure 5-7. A modified and vented inert-gas tungsten arc gun and nozzle are used, with the nozzle pressed against one of the two pieces of the joint. The workpieces must be sufficiently rigid to sustain the pressure that must be applied to one side in order to stayin reasonably good contact. The arc between the tungsten electrode and the upper workpiece provides the necessary heat, and an inert gas, usually argon or helium, flows through the nozzle and provides a shielding atmosphere. Automatic controls move the electrode to make momentary contact with the workpiece to start the arc, and then they withdraw and hold it at a correct distance to maintain the arc. The duration of the arc is timed automatically so that the two workpieces are heated sufficiently to form a spot weld under the pressure of the arc gun's nozzle. The depth and size of the weld nugget are controlled by the amperage, time, and type of shielding gas, as shown in Figure 5-8.

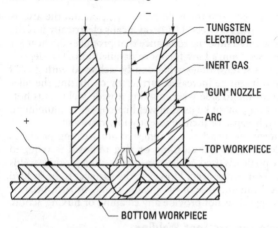

Fig. 5-7 Schematic diagram for making spot welds by the inert-gas-shielded tungsten arc process.

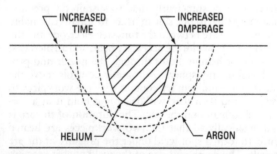

Fig. 5-8 Changes in current, time, and shielding gas affect the shape of the weld nugget.

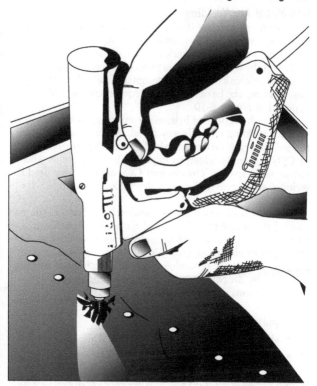

Fig. 5-9 Making a spot weld by the inert-gas-shielded tungsten arc process.

Because access to only one side of the work is required, this type of spot welding has an advantage over resistance spot welding in certain applications, such as in fastening relatively thin sheet metal to a heavier framework (see Figure 5-9).

Gas Metal Arc Welding

Gas metal arc welding (GMAW) was a logical outgrowth of gas tungsten arc welding, differing in that the arc is maintained between an automatically fed, consumable wire electrode and the workpiece. This process provides the additional filler automatically. Formerly designated as MIG (metal, inert gas) welding, the basic circuit and equipment for GMAW are depicted in Figure 5-10.

Although argon and helium, or mixtures of them, can be used for welding virtually any metal, they are used primarily for welding nonferrous metals. For welding steel, some O_2 or CO_2 is usually added to improve the arc stability and to reduce weld spatter. The cheaper CO_2 alone can be used for welding steel, provided that a deoxidizing electrode wire is employed.

The shielding gases have considerable effect on the nature of the metal transfer (drop) from the electrode to the work, and they also affect the tendency for undercutting. Several types of electronic controls that alter the wave form of the current make it possible to vary the mechanism of metal transfer—by drops, spray, or short-circuiting drops. Some of these variations in the basic process are:

- pulsed arc welding (GNAW-P)
- short-circuiting arc welding (GNAWS)
- spray-transfer welding (GMAW-ST)

Buried-arc welding (GMAW-B) is another variation, in which carbon dioxide-rich gas is used and the arc is buried in its own crater. Figure 5-11 illustrates one system for producing a series of peak current surges that preheat the electrode just before droplet formation.

Gas metal arc welding is fast and economical because there is no frequent changing of electrodes, as when welding with stick-type electrodes. In addition, there is no slag formed over the weld; the process often can be automated; and, if done

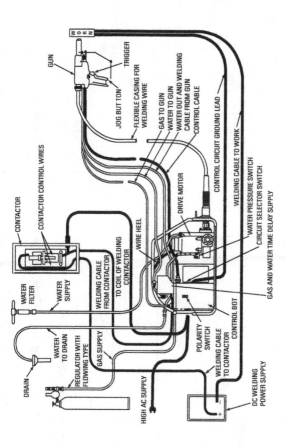

Fig. 5-10 Circuit diagram for GMAW welding.

131

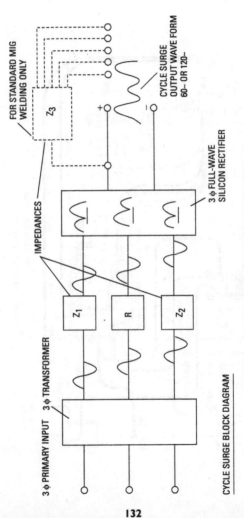

Fig. 5-11 Cycle surge block diagram for MIG welding.

FOR STANDARD MIG WELDING ONLY

CYCLE SURGE OUTPUT WAVE FORM 60∼ OR 120∼

Z_3

IMPEDANCES

3 φ FULL-WAVE SILICON RECTIFIER

Z_1 R Z_2

3 φ PRIMARY INPUT 3 φ TRANSFORMER

CYCLE SURGE BLOCK DIAGRAM

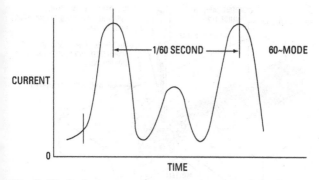

Fig. 5-12 Peak-current surges are used to control droplet transfer.

manually, the welding head is relatively light and compact, as shown in Figure 5-12.

FLUX CORED ARC WELDING

Flux cored arc welding (FCAW) utilizes a continuous electrode wire that contains a granular flux within its hollow core. Additional shielding is often provided by a small flow of CO_2 around the arc zone. This is done so often it is considered to be a variation of gas metal arc welding. FCAW is used primarily for welding steel, and the basic arrangement is shown in Figure 5-13. Because the equipment is larger, it is not as readily portable or as easily manipulated as regular GMAW equipment. Excellent welds can be made with the FCAW process, however.

SUBMERGED ARC WELDING

In submerged arc welding (SAW), as shown in Figure 5-14, the arc is maintained beneath a granular flux. Either AC or DC current can be used as the power source. The flux is deposited just ahead of the electrode, which is in the form of coiled

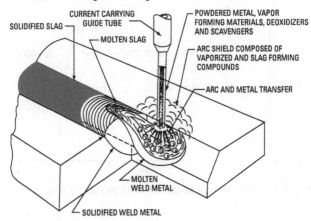

Fig. 5-13 Schematic representation of the flux-cored arc welding process (FCAW).

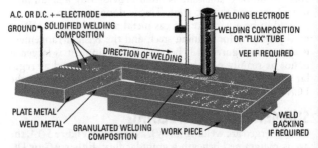

Fig. 5-14 Basic features of the submerged arc welding process (SAW).

wire that is copper-coated to provide good electrical contact. Because the arc is completely submerged in the flux, only a few small flames are visible.

Granular flux provides excellent shielding of the molten metal. Because the pool of molten metal is relatively large, good fluxing action occurs and removes impurities. This produces welds of very high quality. A portion of the flux is melted and solidifies into a glasslike covering over the weld. Along with the flux that is not melted, this covering provides a thermal coating that slows down the cooling of the weld area and thus helps to produce a soft, ductile weld. The solidified flux cracks loose from the weld on cooling and is easily removed. Surplus unmelted flux can be recovered with a vacuum cleaner.

Submerged arc welding is most suitable for making flat butt or fillet welds. The equipment used for SAW varies. The portable equipment shown in Figure 5-15 moves the welding head along the work automatically. Production setups are shown in Figure 5-16, and manual equipment in Figure 5-17. In the latter, the electrode wire is fed through a funnel device that also contains the flux and the electrode guide. The arc length is controlled automatically through the feeding of the wire. Such equipment is very useful for making short or non-linear welds in cases where the time required for setting up the usual equipment would not be justified.

High welding speeds and welds in thick plates can be achieved with the ordinary submerged arc process. Welding speeds of 762 mm (30 inches) per minute in 25.4-mm (1-inch) steel plate, or 304.8 mm (12 inches) per minute in 38.1-mm (1½-inch) plate, are common. Because the metal is deposited in fewer passes than are needed in manual arc welding, there is less probability of entrapped slag or voids in the weld, and higher-quality welds are more consistently obtained.

Welding heads with a single electrode are most commonly used, but there are machines that use three electrodes and operate from three-phase alternating current. Three-phase

Fig. 5-15 Making a submerged arc weld using portable equipment.

power is often employed in production work. The SAW process is widely used for large-volume welding, as in the building of ships or the manufacture of large-diameter steel pipes or tanks.

A modification of the basic process uses iron powder deposited in the prepared gap between the plates to be joined. It is deposited ahead of the flux and in conjunction with a backing strip. As with the iron powder-coated electrodes used in stick welding, this iron powder process permits substantially higher metal deposition rates to be obtained.

The arrangement shown in Figure 5-17 allows vertical welds to be made by the submerged arc process. Stationary

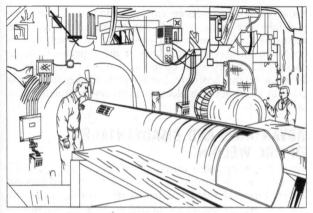

Fig. 5-16 Production setup for welding large pipe by the submerged arc process.

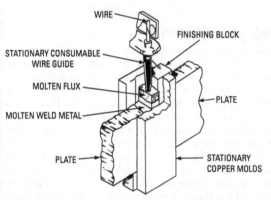

Fig. 5-17 Setup for making vertical welds by the submerged arc process.

molds with copper sides are employed, and a stationary, consumable wire guide is used. The consumable wire guide, coated with flux, melts as it enters the flux pool, thereby replacing the flux that solidifies at the copper-weld interface. Good quality welds up to 102 mm (4 inches) thick can be made by this procdure, but for plate thicker than about 51 mm (2 inches) the electroslag process is usually more economical.

ADVANTAGES AND DISADVANTAGES OF ARC WELDING

Great flexibility and a variety of processes make arc welding an extremely useful, versatile, and widely used process. Except for gas-shielded tungsten arc spot welding, stud welding, and to some degree submerged arc welding, the various arc welding processes have one disadvantage—the quality of the weld depends ultimately on the skill and integrity of the worker who does the welding. It is easy to see that proper training, selection, and supervision of personnel are of even greater importance than adequate inspection.

PLASMA ARC WELDING

Plasma welding is a very-high-temperature welding process that uses an electric arc to energize the gas in a confined chamber. This produces a plasma jet that is used to melt metals. The ionized argon or helium gas is heated with so much energy that its physical state changes into plasma. Plasma is a different form of matter. The plasma gas is in an unstable condition and has a great amount of energy, so it passes out of a restricted orifice to produce a plasma jet traveling at sonic speeds. When it strikes the workpiece, the plasma gas goes back to its natural state, thereby releasing the heat energy that was stored in the plasma. The temperatures produced by this release of the plasma's energy range between 500°F and 60,000°F. This is far above normal welding temperatures in the 6500°F range. See Figure 5-18.

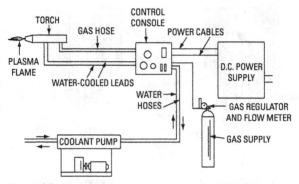

Fig. 5-18 A plasma welding system needs to control a large amount of electrical power.

Plasma Arc Torches

There are two types of plasma arc torches (see Figure 5-19).

- The transferred type, for welding and cutting
- The nontransferred type, for welding nonconductive materials and for metallizing and hard surfacing

Plasma arc torches may be operated manually or automatically. The welding operations performed may be the same as in gas tungsten arc welding, includes edge welds and butt welds. However, the additional amount of heat energy it is sufficient for keyhole welding as well.

Keyhole Welding

Keyhole welding is accomplished by the plasma melting a hole completely through the metal being welded (see Figure 5-19). The edges of the hole are liquid, and the metal flows around the edge of the hole and the plasma flame as the torch is moved forward. This liquid metal meets at the rear of the hole, where surface tension causes it to flow together and

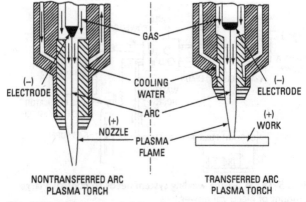

Fig. 5-19 Two types of plasma arcs are the nontransferred arcs used for nonconductive materials and the transferred arcs used for welding and cutting metals.

produce a weld. This type of welding is commonly done with a square butt joint on:

- carbon or low-alloy steels
- stainless steels
- titanium
- copper
- brass

Advantages of plasma welding: Higher welding speeds than gas tungsten arc welding in many applications

- Produces cleaner welds without tungsten contamination in the weld
- Requires less skill from the operator
- Has the ability to produce products at a more competitive market price

A plasma arc weld schematic is shown in Figure 5-20.

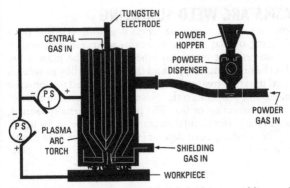

Fig. 5-20 (a) Schematic drawing of plasma arc welding surface torch, powder connections, and powder dispenser.

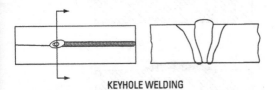

KEYHOLE WELDING

Fig. 5-20 (b) Keyhole welding melts through the workpiece as the hole is moved across the metal. The liquid steel edges flow together and cool as one piece.

PLASMA ARC WELD SURFACING

Plasma arc weld surfacing applies a thin coating or a hard facing overlay of heat-resistant, wear-resistant, and/or corrosion-resistant metals to other metals. The plasma arc used is the nontransfer type, and the metal or alloy powder is metered into the stream of argon gas. It is carried to a powder chamber and placed into the plasma stream beneath the upper arc-constructing orifice. The powder is partially melted, heated in the plasma, and then completely melted in the liquid pool. The resulting hard deposit is welded to the base metal. See Figure 5-21.

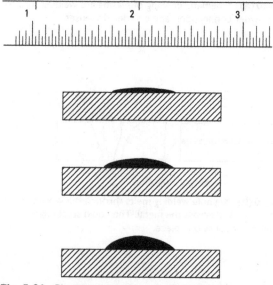

Fig. 5-21   Plasma arc overlays of cobalt surfacing.

Applications of Hard Surfacing

Hard surfacing is used in a number of applications, mostly in the construction industry. Among them are:

- Ditch digging teeth
- Plow points
- Bulldozer blade edges
- Turbine blades
- Extruder screw flights
- Anything that needs to be tough enough to withstand abrasive wear

PLASMA ARC CUTTING

Plasma arc cutting is used to cut various metals at high speeds (see Figure 5-22):

- nonferrous metals
- stainless steels
- refractories
- carbon steels

Carbon steel up to 2 inches thick can be cut up to 10 times faster by plasma arc cutting than by oxyacetylene flame. A 1/4-inch carbon steel plate can be cut at up to 200 inches per minute, with quality equal to a cut by oxyacetylene flame. See Figure 5-23.

Stainless steel up to 2 inches thick can be cut with gas mixtures of argon and hydrogen. Most nonferrous metals can also be cut quickly. Aluminum up to 5 inches thick can be cut, and 1/4-inch aluminum may be cut at the rate of 300 inches per minute. This method of cutting of aluminum is a welcome innovation, because prior to plasma arc cutting aluminum had to be cut by mechanical means.

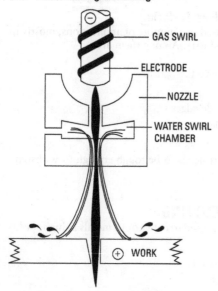

Fig. 5-22 A water-constricted plasma arc concentrates the plasma jet into a narrow cutting stream.

Adding an oxygen jet to the plasma arc cutting process can increase the cutting action on steels without causing erosion to the tungsten cathode. The addition of oxygen to the cutting system has enabled steels to be cut more rapidly, which makes the cost of large cutting contracts more economical. Initial investment for the equipment is high. However, the production rate is such that the cost savings ratio compared to oxyacetylene cutting is about three to one if large amounts of cutting are done.

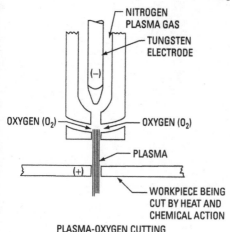

NITROGEN
PLASMA GAS

TUNGSTEN
ELECTRODE

(−)

OXYGEN (O$_2$) OXYGEN (O$_2$)

PLASMA

(+)

WORKPIECE BEING
CUT BY HEAT AND
CHEMICAL ACTION

PLASMA-OXYGEN CUTTING

Fig. 5-23 Plasma-oxygen cutting increases the speed of carbon steel cutting.

Advantages of plasma welding: Higher welding speeds than gas tungsten arc welding in many applications

- Produces cleaner welds without tungsten contamination in the weld
- Requires less skill from the operator
- Has the ability to produce products at a more competitive market price

6. BRAZING AND BRAZE WELDING

Brazing, also known as hard soldering, is similar to bronze welding. The bonding of metals together with a bronze metal that melts above 800°F, but does not melt the base metal, is bronze welding. Metal clearances and the bead are similar to those in the oxyacetylene welding of steel, but a large amount of filler metal is used.

Brazing differs from other welding processes in that a non-ferrous filler metal is used that melts at a temperature above 450°C (840°F). This is below the solidus temperature of the base metal or metals. Coalescence is between the filler and the base metals, or between the base metals and alloys that may be formed. As a result, the base metals are not melted, and capillary attraction plays an important role in distributing the filler metal in the joint.

Some important characteristics and requirements of brazing are:

- The 450°C (840°F) temperature is an arbitrary one, and is used to distinguish brazing from soldering.

- Because the base metals are not melted, the brazing metal always is different in composition from the base metals.

- For capillary attraction to exist, the clearance between the parts being joined must be quite small.

- In order for capillary attraction to be effective, the base metal surfaces must be clean so that the brazing metal can easily wet them. This usually requires the use of a flux, or inert atmospheres in a vacuum furnace.

- Heating a brazed joint above the melting point of the brazing metal, even with subsequent cooling, may destroy the integrity of the joint.

BRAZING ADVANTAGES

Virtually all metals can be joined by some type of brazing metal; coalescence is always between the base metal and a dissimilar brazing metal.

- Less heating is required than for welding
- The process can be done more quickly and more economically
- Lower temperatures are used
- Fewer difficulties due to distortion are encountered
- Thinner and more complex assemblies can be joined successfully
- Many brazing operations can be mechanized
- Coalescense is the growing together or growth into one body of the base metal parts.

These advantages make it apparent that brazing is well suited for use in mass production and for delicate assemblies.

Disadvantage of brazing

A disadvantage of brazing is the fact that reheating can cause inadvertent melting of the braze metal, causing it to run out of the joint, thus weakening or destroying the joint.

Too often this occurs when people apply heat to brazed parts in attempting to repair or straighten such devices as bicycles or motorcycles. Such a consequence, of course, is not a defect of brazing, but it can lead to most unfortunate results. When brazing is specified for use in products that later might be subjected to abuse, adequate warning should be given to those who use the device.

NATURE AND STRENGTH OF BRAZED JOINTS

A brazed joint obtains its strength from a combination of the braze metal and the base-metal alloy that is formed. Penetration of the low-viscosity brazing metal into the grain boundaries of the base metal also adds strength. The strength of a properly brazed joint is between that of the base metal and the braze metal. However, the strength also is a function of the clearance between the parts being brazed. There must be sufficient clearance so that the braze metal will wet the joint and flow into it. Beyond this amount of clearance, the strength decreases rapidly to that of the braze metal. The proper amount of clearance varies considerably, depending primarily on the type of braze metal. Copper requires virtually no clearance when heated in a hydrogen atmosphere.

A slight press fit of about 0.1 per cent per unit of diameter (0.001-inch per inch) is recommended. For silver-alloy brazing metals the proper absolute clearance is about 0.04 to 0.05 mm (0.0015 to 0.002 inch).

When 60-40 brass is used for brazing iron and copper, a clearance of about 0.52 to 0.76 mm (0.02 to 0.03 inch is desirable).

BRAZING METALS
The most commonly used brazing metals are:

- brazing brass
- copper silicon
- copper phosphorus
- nickel silver
- aluminum alloys
- silver alloys
- manganese bronze

Table 6-1 lists some of the more frequently used brazing metals and their uses. Table 6-2 shows welding tip sizes and the gas pressures most often used.

Copper is used only for brazing steel and other high-melting-point alloys, such as high-speed steel and tungsten carbide. Its use is confined almost exclusively to furnace

Table 6-1   Brazing Metals and Their Uses

Braze Metal	Composition	Brazing Process	Base Metals
Brazing brass	60% Cu, 40% Zn	Torch Furnace Dip Flow	Steel, copper, high copper alloys, nickel, nickel alloys, stainless steel
Manganese bronze	58.5% Cu, 1% Sn, 1% Fe, 0.25% Mn, 39.5% Zn	Torch	Steel, copper, high copper alloys, nickel, nickel alloys, stainless steel
Nickel silver	18% Ni, 55–65% Cu, 27–17% Zn	Torch Induction	Steel, nickel, nickel alloys
Copper silicon	1.5% Si, 0.25% Mn, 98.25% Cu, 1.5% Si, 1.00% Zn, 97.5% Cu	Torch	Steel
Silver alloys (no phosphorus)	5–80% Ag, 15–52% Cu, balance Zn + Sn + Cd	Torch Furnace Induction Resistance Dip	Steel, copper, copper alloys, nickel, nickel alloys, stainless steel
Silver alloys (with phosphorus)	15% Ag, 5% P, 80% Cu	Torch Furnace Induction Resistance Dip	Copper, copper alloys
Copper phosphorus	93% Cu, 7% P	Torch Furnace Induction Resistance	Copper, copper alloys

**Table 6-2 Welding Tip Sizes and Pressure Settings
For Brazing**

Metal Thickness (inches)	Tip Size (inches)	Size of Brazing Rod (inches)	Oxygen Pressure (psi)	Acetylene Pressure (psi)
1/32	1	1/16	5	5
3/64	2	1/16	5	5
1/16	3	1/16	5	5
3/32	4	3/32	5	5
1/8	5	3/32	5	5
3/16	6	3/32	6	6
1/4	7	1/8	7	7
5/16	8	5/32	8	8

heating in a protective hydrogen atmosphere in which the copper is extremely fluid and requires no flux.

Copper brazing is used extensively for assemblies composed of low-carbon steel stampings, screw-machine parts, and tubing, such as are common in mass-produced products. Copper alloys, in the form of "spelter" brass, were the earliest brazing materials. The most common use today for copper alloys in brazing is:

- In the repair of steel and iron castings.
- Tobin and manganese bronzes are frequently used for this purpose.
- Phos-copper is used extensively in brazing copper (should not be used on ferrous alloys).
- Silver alloys are widely used in fabricating copper and nickel alloys. Although these brazing alloys are expensive, such a small amount is required that the cost per joint is low.

One typical silver-phosphorus-copper alloy, known as Sil-Fos, is self-fluxing on clean copper and is used extensively for brazing this material.

Silver alloys also are used for brazing stainless steel. However, because the brazing temperatures are in the range of carbide precipitation, only stabilized stainless steels should be brazed with these alloys if continued corrosion resistance is desired.

Aluminum silicon alloys, containing about 6 to 12 per cent silicon, are used for brazing aluminum and aluminum alloys. By using a braze metal that is not greatly unlike the base metal, the possibility of galvanic corrosion is reduced. However, because these brazing alloys have melting points of about 610°C (1130°F), when the melting temperature of commonly brazed aluminum alloys, such as 3003, is around 669°C (1290°F), control of the temperature used in brazing is quite critical. Thus, in brazing aluminum, good temperature control must be exercised, and proper fluxing action, surface cleaning, and/or use of a controlled-atmosphere or vacuum environment must be utilized to assure adequate flow of the braze metal and yet avoid damage to the base metal.

A commonly used procedure in connection with brazing aluminum is to use sheets that have one or both surfaces coated with the brazing alloy. The thickness of the coat is about 10 per cent of the total sheet thickness. These "brazing sheets" have sufficient coating to form adequate fillets. Joints are made merely by coating the joint area with suitable flux followed by heating.

Fluxes

Brazing fluxes play a very important part in brazing by:

(1) dissolving oxides that may be on the surface prior to heating

(2) preventing the formation of oxides during heating

(3) lowering the surface tension of the molten brazing metal, which promotes its flow into the joint

Cleanliness is the primary factor affecting the quality and uniformity of brazed joints. Although fluxes will dissolve modest amounts of oxides, they are not cleaners. Before a flux is applied, all dirt—particularly oil—should be removed from the surfaces that are to be brazed. The less the flux has to do prior to heating, the more effective it will be during heating.

Borax was a commonly used brazing flux at one time. Fused borax should be used because the water in ordinary borax causes bubbling when heat is applied to the flux. Alcohol can be mixed with fused borax to form a paste.

Many modern fluxes are available that have lower melting temperatures than borax and are somewhat more effective in removing oxidation. Fluxes should be selected with reference to the base metal. Paste fluxes usually are used for furnace, induction, and dip brazing and either paste or powdered fluxes are used for torch brazing. In furnace, induction, and dip brazing the flux ordinarily is brushed onto the surfaces. In torch brazing it is often applied by dipping the heated end of the filler wire into the flux.

Fluxes for aluminum usually are mixtures of metallic halide salts. The base typically is potassium chloride (from 15 to 85 per cent of the flux). Activators, such as fluorides or lithium compounds, are added. These fluxes do not dissolve the surface oxide film on aluminum.

Most brazing fluxes are corrosive. The residue should be removed from the work after the brazing is completed. This is particularly important in the case of aluminum. Much effort has been devoted to developing fluxless procedures for brazing aluminum.

Applying Brazing Metal

Brazing metal is applied to joints in three ways. In the oldest method, commonly used in torch brazing, the brazing metal is in the form of a rod or wire. When the joint has been heated to a sufficient temperature so that the base metal will melt, the brazing wire or rod is then melted by the torch and capillary attraction causes it to flow into the joint. Although the base metal should be hot enough to melt the braze metal, and assure its remaining molten and flowing into the joint, the actual melting should be done with the torch.

Obviously, this method of braze metal application requires considerable labor. Since the brazing metal is always applied from the outside, care is necessary to ensure that it has flowed to the inner portions of the joint. To avoid these difficulties, the braze metal often is applied to the joint before heating, in the form of wire or shims. In cases where it can be done, rings or shims of braze metal often are fitted into internal grooves in the joint before the parts are assembled, as shown in Figure 6-1. When this procedure is employed, the parts

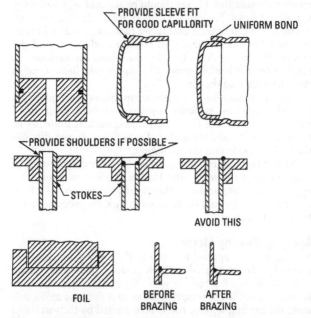

Fig. 6-1 Applying braze metal in sheet or wire form to assure flow into the joint.

usually must be held together by press fits, riveting, staking, tack welding, or a jig, to assure their proper alignment. In such preloaded joints, care must be exercised to assure that the filler metal is not pulled away from the intended surface by the capillary action of another surface with which it may be in contact. Capillary action always will pull the molten braze metal into the smallest clearance, whether or not such was intended.

Another precaution that must be observed is that the flow of filler metal not be cut off by the absence of required clearances or by no provision for the escape of trapped air. Also, fillets or grooves within the joint may act as reservoirs and trap the metal.

Special brazing jigs or fixtures often are used to hold the parts that are to be brazed in proper relationship during heating, especially in the case of complex assemblies. When these are used, it usually is necessary to provide springs that will compensate for expansion, particularly when two or more dissimilar metals are being joined. Figure 6-2 shows an excellent example of this procedure.

HEATING METHODS USED IN BRAZING

A common source of heat for brazing is a gas-flame torch. In this torch-brazing procedure, oxyacetylene, oxyhydrogen, or other gas-flame sources can be used. Most repair brazing is done in this manner because of its flexibility and simplicity, but the process also is widely used in production brazing, Its major drawbacks are the difficulty in obtaining uniform heating, proper control of the temperature, and the requirement of costly skilled labor. In production-type torch brazing, specially shaped torches often are used to speed the heating and to aid in reducing the amount of skill required.

Large amounts of brazing are done in controlled-atmosphere furnaces. In such furnace brazing, the brazing metal must be preloaded into the work. If the work is not of such a nature that its preassembly will hold the parts in

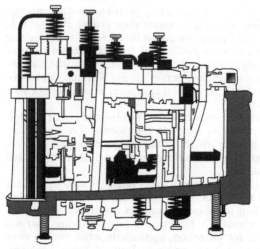

Fig. 6-2 Brazing fixture with provision for heat-caused expansion.

proper alignment and with adequate pressure, brazing jigs or fixtures must be used. Assemblies that are to be brazed usually can be designed so that such jigs or fixtures will not be needed if adequate consideration is given to this matter; often a light press fit will suffice.

Figure 6-3 shows a number of typical furnace-brazed assemblies. Because excellent control of brazing temperatures can be obtained and no skilled labor is required, furnace brazing is particularly well suited for mass production. Either box- or continuous-type furnaces can be used, the latter being more suitable for mass-production work. Furnace brazing is very economical, particularly when no flux is required, as in brazing steel parts with copper.

In salt-bath brazing the parts are heated by dipping in a bath of molten salt that is maintained at a temperature slightly

Fig. 6-3 Furnace-brazed assemblies.

above the melting point of the brazing metal. This method has three major advantages: (1) the work heats very rapidly because it is completely in contact with the heating medium, (2) the salt bath acts as a protective medium to prevent oxidation, and (3) thin pieces can easily be attached to thicker pieces without danger of overheating because the temperature of the salt bath is below the melting point of the parent metal; this latter feature makes this process well suited for brazing aluminum.

It is essential that parts that are to be dip-brazed be held in jigs or fixtures, or that they be prefastened in some manner and that the brazing metal be preloaded into the work. Also, to assure that the bath remains at the desired temperature, the volume must be rather large, depending on the weight and quantity of the assemblies that are to be brazed.

In dip brazing, the assemblies are immersed in a bath of molten brazing metal. The bath thus provides both the required heat and the braze metal for the joint. Because the braze metal usually will coat the entire work, it is wasteful of braze metal and is used primarily only for small parts, such as wire. Induction brazing utilizes high-frequency induction currents for heating. The process has the following advantages, which account for its extensive use:

- Heating is very rapid—usually only a few seconds being required for the complete cycle.

- The operation can easily be made semiautomatic, so that semiskilled labor may be used.

- Heating can be confined to the localized area of the joint, because of the shape of the heating coils and the short heating time. This reduces scale, discoloration, and distortion.

- Uniform results are readily obtained.

- By making new, and relatively simple, heating coils, a wide variety of work can be done with a single power unit.

The high-frequency power-supply units are available in both small and large capacities at very modest costs. The only other special equipment required for adapting induction brazing to a given job is a simple heating coil to fit around the joint to provide heating at the desired area. These coils are formed of copper tubing through which cooling

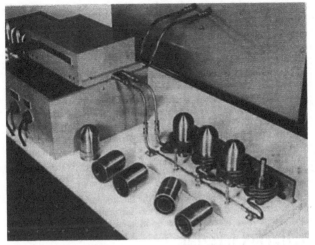

Fig. 6-4 Typical induction brazing operation.

water is carried. Although the filler material can be added to the joint manually after it is heated, the usual practice is to use preloaded joints to speed the operation and obtain more uniform joints. Figure 6-4 shows a typical induction-brazing operation.

Induction brazing is so rapid that it can often be used to braze parts with high surface finishes, such as silver plating, without affecting the finish. Some resistance brazing is done, in which the parts to be joined are held under pressure between two electrodes, similar to spot welding, and an electrical current is passed through them. However, unlike resistance welding, most of the resistance is provided by the electrodes, which are made of carbon or graphite. Most of the heating of the metal thus is by conduction from the hot electrodes. This process is used primarily in the electrical equipment manufacturing industry for brazing conductors, cable connections, and so on. Regular resistance welders and their timing equipment often are adapted for resistance brazing.

FLUX REMOVAL

Although not all brazing fluxes are corrosive, most of them are. Consequently, flux residues usually must be completely removed. Most of the commonly used fluxes are soluble in hot water, so their removal is not difficult. In most cases, immersion in a tank of hot water for a few minutes will give satisfactory results, provided that the water is kept really hot. Usually, it is better to remove the flux residue while it is still hot. Blasting with sand or grit is also an effective method of flux removal, but this procedure cannot be used if the surface finish must be maintained. Such drastic treatment seldom is necessary.

FLUXLESS BRAZING

Both the application of brazing flux and the removal of flux residues involve significant costs, particularly where complex joints and assemblies are involved. Consequently, a large amount of work has been devoted to the development of procedures whereby flux is not required, particularly for brazing aluminum. This work has been spurred by the obvious advantages of aluminum as a lightweight and highly efficient heat conductor for use in heat transfer applications such as automobile radiators, where weight reduction is of increasing importance.

The brazing of aluminum is complicated by the metal's high refractory oxide surface film, its low melting point, and its high galvanic potential. However, the successful fluxless brazing of aluminum has been achieved by employing rather complicated vacuum furnace techniques that use vacuums up to 0.0013 Pa $(1 \times 10^{-5}$ torr). Often a *getter* metal is employed to aid in absorbing the small amount of oxygen, nitrogen, and other occluded gases that remain in the vacuum or may have evolved from the aluminum being brazed. The aluminum must be carefully degreased prior to brazing, and the design of the joint is quite critical. Sharp, V-edge joints appear to give the best results.

Some success has been achieved with the fluxless induction brazing of aluminum in air. The induction heating coils are designed to act as clamps to hold the pieces being brazed together under pressure, and an aluminum braze metal containing about 7 percent silicon and 2.5 percent magnesium is used. The resulting magnesium vapor apparently reduces some of the oxide on the surface of the aluminum, permitting the braze metal to flow and cover the aluminum surface.

DESIGN OF BRAZED JOINTS

Three types of brazed joints are used:

- butt
- scarf
- lap, or shear

These joints and some examples of good and poor design details are shown in Figure 6-5.

Because the basic strength of a brazed joint is somewhat less than that of the parent metals, desired strength must be obtained by using sufficient joint area. This means some type of lap joint when maximum strength is required. If joints are made very carefully, a lap of 1 or $1\frac{1}{4}$ times the thickness of the metal can develop strength as great as that of the parent metal. However, for joints made in routine production it is best to use a lap equal to three times the material thickness. Full electrical conductivity usually can be obtained with a lap about $1\frac{1}{2}$ times the material thickness.

If maximum joint strength is desired, it is important to have some pressure applied to the parts during heating and until the braze metal has cooled sufficiently to attain most

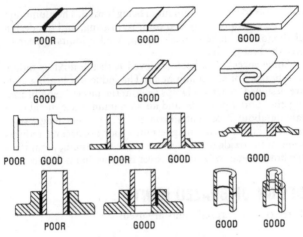

Fig. 6-5 Good and poor joint design for brazing.

of its strength. In many cases, the needed pressure can be obtained automatically through proper joint selection and design.

In designing joints to be brazed, one must make sure that no gases can be trapped within the joint. Trapped gas may prevent the filler metal from flowing throughout the joint because of the pressure that develops during heating.

BRAZE WELDING

Braze welding differs from brazing in that capillary attraction is not used to distribute the filler metal—the molten filler metal is deposited by gravity. Because relatively low temperatures are required, there is less danger of warping than if welding were used. Braze welding is very effective for the repair of steel parts and ferrous castings and is used almost exclusively

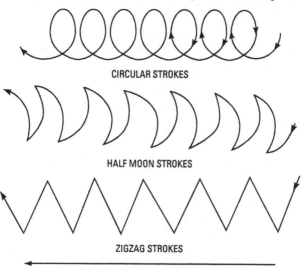

CIRCULAR STROKES

HALF MOON STROKES

ZIGZAG STROKES

DIRECTION OF STROKES

Fig. 6-6 Patterns for strokes to keep metal flowing and parent metal evenly heated.

for this type of work. Obviously, the strength obtained is determined by the braze metal used, and considerable buildup of the braze metal is required if full strength of the repaired part is needed.

Virtually all braze welding is done with an oxyacetylene torch. The surfaces should be tinned with a thin coating of the brazing metal before the remainder of the filler metal is added.

See Figure 6-6 for the strokes used to heat parent metals with the torch. This method can create a better joint, because both pieces are heated equally.

CIRCULAR STROKES

HALF-MOON STROKES

STRAIGHT STROKES

DIRECTION OF STROKES

Fig. 6-6. Exercises for strengthening hand movement flowing and muscle-memory freehand.

for this type of work. Frequently the graphic method is the expression of the ideas of an end user and consequently the field of the ideas involved in their strength of the repeated parts is critical.

Usually all freehanding is done with an overwhelming circle. The balloon should be drawn with a thin casting of the medium used before the examination of the filling up is undertaken.

See Figure 6-6 for the strokes used in hand penmanship in this drawing. This method is accurate and better than just letting the pencil stroke and complete.

7. SOLDERING

Soldering is a low-temperature joining process used to fasten two metal surfaces together without melting either of the metals. Fusion does not take place. A nonferrous filler metal called *solder* or *soft solder* is added to the joint, where it melts at temperatures below 800°F and flows into and along the joint surface by capillary action. This joining process is also called *soft soldering* to distinguish it from *brazing* (sometimes referred to as *hard soldering* or *silver soldering*).

SOLDERING APPLICATIONS

Soldering is a useful process for joining most metals in a broad range of applications (see Table 7-1). For example, it is widely used in the electronics industry, where its low temperatures reduce the possibility of damage to semiconductors and similar components. It is also widely used in HVAC or plumbing work to seal pipes and tubing.

The soldering processes include dip soldering, resistance soldering, wave soldering, infrared soldering, furnace soldering, iron soldering, and torch soldering. Iron soldering and torch soldering, the two common manual processes, are described in this chapter.

Soldering Advantages:

- A self-contained and portable joining process.
- Easy to learn.
- Inexpensive equipment.
- Most metals can be soldered.
- Low temperatures minimize the possibility of damage to components being joined.
- Rapid assembly of parts.
- Mistakes are easily corrected.
- Little or no finishing required.

Table 7-1 Relative Solderability of Common Metals,
Metal Alloys, and Coatings

Base Metal, Alloy, or Applied Coating	Soldering Recommended	Soldering Not Recommended
Aluminum	X	
Aluminum-Bronze	X	
Beryllium		X
Beryllium-Copper	X	
Brass	X	
Cadmium	X	
Cast Iron	X	
Chromium		X
Copper	X	
Copper-Chromium	X	
Copper-Nickel	X	
Copper-Silicon	X	
Gold	X	
Inconel	X	
Lead	X	
Magnesium	X	
Manganese-Bronze (high-tensile)		X
Monel	X	
Nickel	X	
Nichrome	X	
Palladium	X	
Platinum	X	
Rhodium	X	
Silver	X	
Stainless Steel	X	
Steel	X	

(continued)

Table 7-1 (continued)

Base Metal, Alloy, or Applied Coating	Soldering Recommended	Soldering Not Recommended
Tin	X	
Tin-Bronze	X	
Tin-Lead	X	
Tin-Nickel	X	
Tin-Zinc	X	
Titanium		X
Zinc	X	
Zinc die castings	X	

Courtesy American Welding Society

Soldering Disadvantages:

- Soldered joints are seldom used where strength is required.
- Solder and base metal colors may not match.

SOLDERING EQUIPMENT

Joints can be soldered by hand using either a gas torch or an electric soldering iron. Some commonly used types of soldering equipment are listed in Table 7-2.

SOLDERS

Solders are nonferrous filler metals used to join metal surfaces. These soldering metals are formed from fusible alloys. There are a number of different types of solders, and the type used depends on the composition of the metals being joined. Not all metals and metal alloys can be soldered (see Table 7-1). Among those that will not take a soldered joint are titanium, chromium, high-tensile magnesium-bronze, beryllium, cobalt, and silicon.

Table 7-2 Manual Soldering Equipment

Type	Comments
Gas torch	• The *oxy-fuel gas torch* uses oxygen in combination with natural gas or propane.
	• Acetylene is not recommended, because the oxyacetylene flame is generally too hot for soft soldering.
	• The tip sizes used for soldering will vary depending on the thickness and type of metal being soldered. These tips are often designed for use with other types of surface work as well, such as heating or silver brazing.
Soldering iron	• Consists of a large piece of copper drawn to a point or edge and fastened to an iron rod that has a handle grip.
	• Soldering irons are available in sizes ranging from the small, lightweight types to the large, heavy-duty models found in industrial plants.
	• Soldering irons may be heated either internally or externally. External heating is usually done in ovens (furnaces) heated by a gas flame or other means. Internal methods of heating involve either gas or electricity, with the latter being the most common form.
Pencil iron	• Receives its name from its slim, pencil-like appearance.
	• The type most commonly used in nontorch soldering.
	• Available in a variety of different sizes.

(*continued*)

Table 7-2 *(continued)*

Type	Comments
Hatchet iron	• Enables soldering in areas that are not easily accessible.
	• A wide variety of tips are available.
	• Elements are interchangeable to provide a wide wattage range.
Soldering gun	• A quick-heating tool.
	• Used for repair work or intermittent soldering.
Desoldering tool	• Designed to melt and remove solder by first heating it and then removing the molten solder by suction.
	• A vacuum bulb located above the heating iron sucks the molten solder from the surface.
	• Tips are heat-resistant and replaceable.
	• Tip orifices are available in a number of different sizes.
Iron holder	Designed to safely hold a soldering iron that is not being used.

Classifying Solders

Some classifications of solders are based on the ASTM specification B32-58T. One category receives its designation from the silver content of the solder. For example, a 2.5S solder is one containing 97.5 percent lead and 2.3 to 2.7 percent silver (although 2.5 percent is the *desired* level). These are the so-called lead-silver solders.

The largest classification category is based on tin content (for example, a 5A solder contains 5 percent tin and 95 percent lead). There are three letter suffixes (A, B, and C) that can be added to the numeral that indicates the tin percentage.

These letters indicate the three composition classes:

- A—0.12 percent maximum antimony content allowed for solders containing 35 percent tin or more.
- B—0.5 percent maximum antimony content allowed for solders containing 35 percent tin or more.
- C—The antimony content may not exceed 6 percent of the tin content in the so-called tin-antimony solders (those containing 20 to 40 percent tin and an appropriate percentage of antimony).

By way of illustration, a 40A solder will consist of 40 percent tin, 60 percent lead and a maximum of 0.12 percent antimony. The percentage given for tin is the *desired* percentage; the one for lead is a *nominal* percentage. These amounts will vary to some degree. A 40B solder will contain 40 percent tin, 60 percent lead and a maximum of 0.5 percent antimony. Finally, a 40C solder will contain 40 percent tin, 58 percent lead and a maximum of 2.4 percent antimony (6 percent of the tin content).

Types of Solders

Alloys are added to solder to produce filler metals with certain required characteristics. For example, antimony is added for extra strength (under specified conditions) and increased electrical conductivity. Silver added to lead will improve wetting characteristics, particularly on copper or steel. Cadmium, silver, and zinc combinations produce a solder particularly suitable for joining aluminum or dissimilar metals.

Solders can be divided into a number of different groups according to their composition. The largest group and most commonly used are the tin-lead alloy solders. See Tables 7-3 through 7-12.

Table 7-3 Tin-Lead Solders

ASTM Alloy Grade	Tin%	Lead %	Solidus Temp. (solid)°F	Liquidus Temp. (molten)°F	Comments
5A	5	95	518	594	• The 60–40 (60% tin, 40% lead) and 50–50 (50% tin and 50% lead) solders are general-purpose solders and are very easy to use, even for the beginner.
10A	10	90	514	573	
15A	15	85	361	535	
20A	20	80	361	533	
25A	25	75	361	511	
30A	30	70	361	491	• The strongest tin-lead alloy is one composed of 42% tin and 58% lead (in other words, a 42–58 solder).
35A	35	65	361	477	
40A	40	60	361	460	
45A	45	55	361	441	• Solders having a maximum tin content in the 30% range or less are very difficult to use, even for an experienced worker.
50A	50	50	361	421	
60A	60	40	361	374	
70A	70	30	361	378	

Courtesy American Welding Society

Table 7-4 Tin-Antimony-Lead Solders

ASTM Alloy Grade	Sn%	Pb%	Sb%	Solidus Temp. (solid)°F	Liquidus Temp. (molten)°F	Comments
20C	20	80		361	533	• Antimony added to increase solder strength
25C	25	75		361	511	• More difficult to use than tin-lead solders
30C	30	70		361	491	• Cannot be used to solder aluminum, zinc
35C	35	65		361	477	coated sheets, or other metals containing zinc

Courtesy American Welding Society

172

Table 7-5 Tin-Antimony Solders

ASTM Alloy Grade	Sn%	Sb%	Solidus Temp. (solid)°F	Liquidus Temp. (molten)°F	Comments
95TA	95	5	452	464	• High-strength solder • Difficult to use in vertical position because of very narrow pasty range • Recommended for food handling equipment because it does not contain lead

Courtesy American Welding Society

Table 7-6 Tin-Silver Solder

ASTM Alloy Grade	Sn%	Ag%	Solidus Temp. (solid)°F	Liquidus Temp. (molten)°F	Comments
96 5Ts	95.5	3.5	430	430	• Not suited for general purpose soldering because it is so expensive • Recommended for fine instrument soldering, special tube joining, and similar limited applications • Easy to apply with rosin flux • Similar to tin-antimony solders

Courtesy American Welding Society

Table 7-7 Tin-Zinc Solders

Sn%	Ag%	Solidus Temp. (solid)°F	Liquidus Temp. (molten)°F	Comments
91	9	390	390	• Used for joining aluminum.
80	20	390	518	• The 91-9 tin-zinc solder has excellent flow and wetting characteristics, as well as a high resistance to corrosion.
70	30	390	592	
60	40	390	645	• Liquidus (melting) temperature increases as the zinc content increases.
30	70	390	708	

Courtesy American Welding Society

Table 7-8 Lead-Silver Solders

ASTM Alloy Grade	Pb%	Ag%	Sn %	Solidus Temp. (solid)°F	Liquidus Temp. (molten)°F	Comments
2.5 S	97.5	2.5	—	579	579	• Used where strength at moderately elevated temperatures is required.
5.5 S	94.5	5.5	—	579	689	• Silver added in order to wet steel and copper more easily.
1.5 S	97.5	1.5	1.0	588	588	• Tend to corrode in a humid storage atmosphere.
						• The addition of tin reduces corrosion tendency while increasing the wetting and flow characteristics.

Courtesy American Welding Society

Table 7-9 Cadmium-Silver Solder

Cd%	Ag%	Solidus Temp. (solid)°F	Liquidus Temp. (molten)°F	Comments
95	5	640	740	• Used in applications where service temperatures are higher than permissible with lower-melting-point solders. • Used to join aluminum to itself or to dissimilar metals. • Lead content can be a health hazard if not handled carefully.

Courtesy American Welding Society

Table 7-10 Cadmium-Zinc Solders

Cd%	Zn%	Solidus Temp. (solid)°F	Liquidus Temp. (molten)°F	Comments
82.5	17.5	509	509	• Used for soldering aluminum.
40.0	60.0	509	635	
10.0	90.0	509	750	• Improper use can lead to health problems.

Courtesy American Welding Society

Table 7-11 Zinc-Aluminum Solder

Zn%	Al%	Solidus Temp. (solid)°F	Liquidus Temp. (molten)°F	Comments
95	5	720	720	• Used for soldering aluminum. • It creates joints with high strength and good corrosion resistance.

Courtesy American Welding Society

Table 7-12 Indium Solders

Sn%	In%	Bi%	Pb%	Cd%	Solidus Temp. (solid)°F	Liquidus Temp. (molten)°F	Comments
8.3	19.1	44.7	22.6	5.3	117	117	• Indium content above 25% provides good corrosion resistance to alkaline solutions.
12.0	21.0	49.0	18.0	—	136	146	• 50% tin–50% indium solder used for glass-to-metal and glass-to-glass soldering.
12.8	4.0	48.0	25.6	9.6	142	149	
50.0	50.0	—	—	—	243	260	• Indium solders containing bismuth require the use of acid fluxes or pre-coating (not required if bismuth is not present).
48.0	52.0	—	—	—	243	243	• The same fluxes, heating methods, and soldering techniques used for tin-lead solders also apply to the indium solders.

Table 7-13 Some Typical Fusible Alloys

Sn%	Bi%	Pb%	Cd%	Sb%	Solidus Temp. (solid)°F	Liquidus Temp. (molten)°F	Comments
13.3	50.5	26.7	10.0	—	158	158	• Fusible alloys are filler metals designed to solder at temperatures below that of the tin-lead alloys (i.e., below 351°F).
15.5	52.5	32.0	—	—	203	203	
14.5	48.0	28.5	—	9.0	217	440	
—	55.5	44.5	—	—	255	255	• Used for soldering on or near heat-sensitive surfaces.
43.0	57.0	—	—	—	281	281	

179

NOTE

A solder containing too much lead makes a weak joint because the lead does not transfuse with brass. A solder containing too much tin becomes brittle.

Solder Temperatures

The correct temperature for soldering must be determined for each particular job, largely by experiment. It varies with the size of the work, the type of work, the solder composition, and the nature of the flux. Soldering temperatures usually range between 500 and 700°F. The operator should not attempt to solder at temperatures close to 800°F, at which point (approximately) the drosses (the scum thrown off from molten metal) become soluble in the solder. When this happens, the solder is said to be "burnt" and will behave badly.

- When soft solder is heated, it starts to melt at a specific temperature. The melting temperature will depend on the type of solder. As additional heat is applied, the temperature increases until all of the solder is melted. The first temperature is called the *solidus* temperature (melting point), and the second temperature is called the *liquidus* temperature (flowing point). The latter temperature is important because the metal must be heated to a temperature higher than the flowing (liquidus) temperature or the solder will not flow.

- The ability of a solder to flow over a metal surface is one determination of the solderability of that metal. However, the flow characteristics will vary from one solder to the next. For example, tin-lead solder of, say, a 5-to-95 composition will flow quite a bit differently from a 95-to-5 tin-silver solder. Only practice will enable an operator to become accustomed to these differences.

- Soft-soldered joints cool and will solidify at their melting point. However, until the joint has cooled to a temperature *below* the melting point, no strain can be applied to the joint, since solder has no strength while it is only partially solidified. Between the flowing point and the melting point is the so-called plastic range, in which there is a mixture of solid and liquid solder.

- A few solders melt and flow at the same temperature. These are referred to as *eutectic* solders. They are suitable for making bead joints, but are not suitable for running into lap joints. It helps to know the melting and flowing points and the plastic range when changing from a solder of known composition to a substitute solder.

SOLDERING FLUXES

Metal surfaces are covered with thin, invisible oxides and other impurities. These must be removed before the solder is applied or it will not adhere to the surface. The removal of these impurities is accomplished by applying a flux.

Note the Following:

- Do *not* skip the cleaning operation that precedes the application of the flux. Although the flux does remove oxides and other impurities, it is not a complete cleaning operation.

- A *flux* is a chemical that not only removes surface contaminants, but also forms a thin film over the surface to prevent contact with the air. Moreover, the flux contributes to the free-flowing characteristic of the solder.

- No one flux can be assigned to any one metal as being peculiarly suited to that metal for all purposes. The nature of the solder often determines the selection of the flux. In electrical work, for example, the best flux is made of pine-amber rosin, because it does not cause corrosion.

- A corrosive flux, such as one containing zinc chloride solution, should be strictly excluded from any electrical work.

- The *Underwriter's Code* permits the use of a flux composed of five parts zinc chloride, four parts alcohol, three parts glycerine, and water. This preparation permits the solder to flow freely and is not highly corrosive.

- Some flux manufacturers add wetting agents to their fluxes, after experimentally determining the amount to be added. Wetting agents are penetrants that cause the solder film to thin out and cover a larger area. Fluxes with wetting agents are not necessary for 50-50 or 60-40 solders, but they are not detrimental. However, it is advisable to use fluxes with wetting agents for the solders with lower tin content. This is particularly necessary for long joints on copper and for all joints on brasses and bronzes.

Fluxes are commonly classified as either corrosive or non-corrosive (see Table 7-14). Some authorities recommend listing mild (slightly corrosive) fluxes as a third (intermediate) class of fluxes. However, because they result in some corrosion, mild fluxes are often grouped with the corrosive fluxes. Using the two-class distinction as a basis, Table 7-15 lists various metals and metal alloys with the recommended flux (i.e., noncorrosive or corrosive) for each.

SOLDERING IRON METHOD

The *soldering iron* method (also referred to as the *soldering copper* method) is a term used to describe soldering done with the copper-tipped soldering iron. This is the oldest manual method of joining metals with solder.

Table 7-14 Flux Classes and Their Characteristics

Flux Class	Comments
Noncorrosive fluxes	• Rosin and rosin alcohol are noncorrosive fluxes.
	• Although these fluxes will not cause corrosion, their fluxing action is very weak.
	• Moreover, the noncorrosive fluxes leave an unsightly brown stain on the surface.
	• This noncorrosive flux residue is not electrically conductive and must be removed after soldering is completed.
Corrosive fluxes	• Zinc chloride is the principal element found in most corrosive fluxes.
	• Other ingredients include ammonium chloride, sodium chloride, hydrochloric acid, hydrofluoric acid, and water.
	• None of these ingredients is present in every corrosive flux, but every corrosive flux does contain mixtures of some kinds of salts and inorganic acids.
	• The corrosive fluxes have a highly active fluxing action, are relatively stable over different temperature ranges, and are very suitable for solders that have the higher melting temperatures. All flux residues must be removed immediately after soldering, or they will corrode the metal.

(continued)

Table 7-14 (continued)

Flux Class	Comments
Intermediate (mild) fluxes	• The so-called mild or intermediate fluxes are slightly corrosive, but weaker in their fluxing action than the corrosive fluxes. The mild fluxes contain organic acids (citric acid, glutamic acid, etc.), whereas the acids found in the corrosive fluxes are inorganic in origin.
	• Mild fluxes are used for quick spot-soldering jobs. Using them for a longer time results in burning or other forms of breakdown caused by their reaction to prolonged heating. The residue of these fluxes can be easily removed with water.

Courtesy American Welding Society

Care and Dressing of Tips

The life expectancy of a copper-tipped soldering iron is commonly reduced by normal use, careless handling, and line voltage fluctuation.

- Most of the problems associated with care and handling can be eliminated simply by being more careful. Learning and following proper work habits is always important. Every soldering station should also have a soldering iron holder. These not only prevent breakage, burned cords, and other mechanical problems; they also reduce the possibility of the operator being burned.

- Improper care and handling probably contributes more to shortening the life expectancy of soldering irons than

Table 7-15 Flux Recommendations for Various Metals and Metal Alloys

Base Metal, Alloy, or Applied Finish	Noncorrosive Flux	Corrosive Flux	Special Flux and/or Solder Required
Aluminum			X
Aluminum-Bronze			X
Beryllium	Soldering Not Recommended		
Beryllium-Copper		X	
Brass	X	X	
Cadmium	X	X	
Cast Iron			X
Chromium	Soldering Not Recommended		
Copper	X	X	
Copper-Chromium		X	
Copper-Nickel		X	
Copper-Silicon		X	
Gold	X		
Inconel			X
Lead	X	X	
Magnesium			X
Manganese-Bronze (High-tensile)	Soldering Not Recommended		
Monel		X	
Nickel		X	
Nichrome			X
Palladium	X		
Platinum	X		
Rhodium		X	
Silver	X	X	

(continued)

Table 7-15   (continued)

Base Metal, Alloy, or Applied Finish	Noncorrosive Flux	Corrosive Flux	Special Flux and/or Solder Required
Stainless Steel			X
Steel	X		
Tin	X	X	
Tin-Bronze	X	X	
Tin-Lead	X	X	
Tin-Nickel	X	X	
Tin-Zinc	X	X	
Titanium	Soldering Not Recommended		
Zinc		X	
Zinc die castings			x

Courtesy American Welding Society

line voltage fluctuations. Some examples of improper handling are: (1) damaged cords (usually by allowing the cord to come into contact with the hot tip, (2) cracked handles (often by dropping), (3) damaged tips (again by dropping), and (4) overheated irons (generally caused by prolonged idling periods).

- The soldering tip will deteriorate over time as a result of normal use. Proper care can prolong its life expectancy.

- Line voltage fluctuations involving current surges or drops are a problem in production work. A surge of current can damage the heating element of the iron. On the other hand, drops in voltage result in reduced heat output that indirectly affects the tip. Line voltage fluctuations can be guarded against by constantly checking the voltage.

Table 7-16 Some Useful Soldering Tips and Techniques

Tip/Technique	
Tinning the tip	• A properly tinned tip is extremely important to a successful soldering operation. It is impossible to solder with an untinned or badly tinned tip, because the oxidized film of copper on the surface prevents the ready transmission of heat.
	• Tinning the tip consists of coating it with solder before beginning the work.
	• The tip is tinned by heating it in a fire or gas flame until hot enough to rapidly melt a stick of solder pressed against it.
	• The solder covering the surface of the tip must be replaced frequently in order to maintain a clean coating.
	• When the tip is heated to the correct temperature, the face of the copper should be cleaned up with an old file. If the tip is too hot, the copper surface will tarnish immediately. To correct this, allow the tip to cool slightly and repeat the cleaning.
	• When the surface slowly begins to tarnish, sprinkle a little flux on it and then rub it with a solder stick.
	• After the molten metal has spread over the surface to be tinned, the superfluous solder is quickly and carefully wiped off with a clean, damp rag.
	• The tip surface should have a bright silver appearance when properly tinned.

(continued)

Table 7-16 *(continued)*

Tip/Technique	
Precoating the surface	• *Precoating* is a term used to describe the covering of a metal surface with a thin, permanent coating of molten solder, pure tin, copper, nickel or some other suitable metal.
	• Surface precoating is sometimes referred to as *tinning*.
	• This term is sometimes incorrectly used to mean *wetting*. Wetting refers to the spreading characteristics of the solder itself.
	• During the precoating procedure, a strong semichemical attraction occurs between the atoms of the precoating metal and those of the base metal. The precoating film, in turn, functions as a base coat for the flux and the second layer of solder.
	• Precoating greatly facilitates the flow of solder on small parts and assemblies.
Wiping	• *Wiping* is a technique used for sealing lead joints.
	• The solders used for wiping contain 30 to 40% tin, 58 to 68% lead, and the remainder antimony.
	• The solder is first melted and then poured in the molten state over the joint.

(continued)

Table 7-16 (continued)

Tip/Technique	
Sweating	• *Sweating* is a term sometimes used to refer to a procedure for temporarily holding together work that has to be turned or shaped—work that could not be so conveniently held by other methods. After the turning or shaping, the parts are readily separated with the aid of heat. • In this operation, the surfaces are cleaned, fluxed, heated, and covered with a film of solder. The soldered surfaces are then placed together and heated by passing the tip over the outside surface until the solder melts and unites the two surfaces. • The procedure for sweating two metal surfaces together is as follows: 1. Clean the two surfaces thoroughly. 2. Apply a flux to both surfaces. 3. Place a piece of tinfoil over one of the surfaces. 4. Place the other surface on top of the tinfoil. 5. Clamp the two surfaces together. 6. Heat the combined surfaces with a hot tip (or a torch if the combined metals have considerable thickness) until solder melts into the joint. 7. Allow the solder to cool until the two surfaces are firmly joined.

Common Problems with Soldering Iron Tips

The three most common tip malfunctions that occur with normal use are tip erosion, tip corrosion, and tip freezing.

- Plain copper tips (i.e., those without special coatings) are those most subject to tip erosion. During the soldering process, small bits of the tip break off and are carried away, which eventually destroys the effectiveness of the tip. Because little can be done about this, plain copper tips are not used for extensive production operations.

- Copper tips with coatings resistant to erosion are designed to prolong the life of the tip. This erosion-resistant coating is a thin plating of metal that can be tinned. The coating is also soluble in solder, but to a much lesser extent than a plain copper tip. Oxides will form on the surface of a coated tip and must be removed. A wire brush is recommended for this purpose, but brushing beyond the exposure of the original plating will ruin the tip. The tip should be retinned immediately after brushing.

- The formation of oxides on the surface of the tip is known as *tip corrosion*. A wire brush or file is recommended for oxide removal. Oxides should be frequently removed, care should be taken not to remove the original coating, and the tip should be immediately retinned.

- While soldering, scale will frequently collect inside the core of the soldering iron. If the scale is not removed, it will solidify and freeze the tip, making it almost impossible to remove from the iron. The chance of this occurring can be reduced by removing the tip after soldering and cleaning off any scale that may have formed inside the core. A frozen tip will frequently result in the destruction of the heating element itself. Consequently, it is a wise move to prevent freezing from occurring in the first place.

Cleaning the Surface

Thoroughly cleaning the surface is perhaps the single most important step in soldering. The solder will not alloy to the surface of the base metal unless it is perfectly clean. Any surface contaminants (dirt, grease, etc.) will obstruct the alloying or wetting process of the solder.

Cleaning can be done mechanically or chemically. Mechanical cleaning may be accomplished by machining, grinding, blasting (with sand, grit, or shot), or hand abrading. Generally, hand abrading is the method most commonly used in the small workshop. It can be done in a number of ways, including: (1) rubbing with (fine grade) steel wool, (2) filing, (3) brushing with a wire brush, or scraping.

Rub the metal surface until it has a bright, shiny appearance. The only exceptions to this rule are a tin surface or one to which a noncorrosive flux is to be applied. Both can be cleaned by wiping with a suitable solvent.

The surface can be chemically cleaned by wiping it with a clean cloth dipped in a solvent or detergent. This will remove most forms of grease, oil, or dirt. The chemical cleaning agent should be removed with hot water before applying the flux.

JOINT DESIGNS FOR SOLDERING

A soldered joint does not have the strength of a welded joint, because the base metal does not melt and fuse with the filler metal. Consequently, a suitable joint design for soldering is determined by the stresses or loads to which it will be subjected. The operator should consider the degree to which the joint must withstand vibration, impact, tension, compression, and shear stress. The more pronounced these factors become, the greater the need for such additional support as bolts, screws, or other fastening devices.

The electrical conductivity of a soldered joint is also an important consideration. Because so many soldered joints serve as electrical connections (e.g., on printed circuit boards), a joint that conducts electricity poorly would hardly be acceptable. Both the composition of the solder and the finished joint must be conducive to a high degree of electrical conductivity. The operator must produce a soldered joint without points and must avoid bridging electrical connections.

The lap and butt joints are the most common joints used for soldering. Of the two, the lap joint is recommended and preferred by most operators. Figure 7-1 illustrates many of the typical joint designs used for soldering.

SOLDERING PROCEDURE

The process of soldering should be carried out as quickly as possible, especially in the case of electrical work. This gives greater protection from damage to the insulation around the wire, because the tendency to burn the insulation is less with a hot tip (a *quick* tip) than with a cooler one.

Although soldering may appear to be easy, it requires a great deal of practice before even a passing skill can be acquired. The following procedure is a basic, step-by-step outline for soldering:

- Before fluxing, remove all dirt and grease from the surfaces to be soldered. This can best be done by wiping the surface with a cloth dipped in an inorganic cleaner or detergent, or by dipping the piece itself in the solution. Finally, clean the surface with fine steel wool until it is bright and free from oxide.

- Follow immediately with a flux.

- Make certain the surfaces fit and contact each other as perfectly as possible.

- Heat the metal—*not* the solder—to a point above the flowing temperature of the solder. Heat the lapped joint

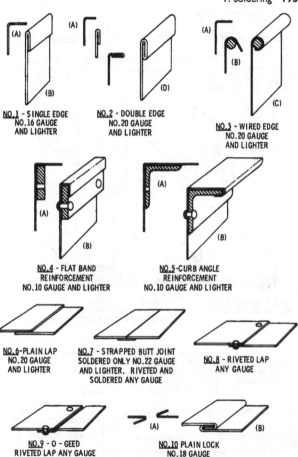

Fig. 7-1 Recommended joint designs for soldering.

(Courtesy American Welding Society)

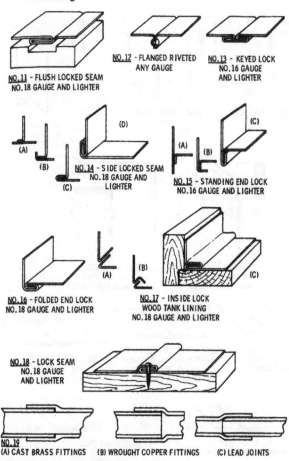

NO. 11 - FLUSH LOCKED SEAM
NO. 18 GAUGE AND LIGHTER

NO. 12 - FLANGED RIVETED
ANY GAUGE

NO. 13 - KEYED LOCK
NO. 16 GAUGE
AND LIGHTER

NO. 14 - SIDE LOCKED SEAM
NO. 18 GAUGE AND
LIGHTER

NO. 15 - STANDING END LOCK
NO. 16 GAUGE AND LIGHTER

NO. 16 - FOLDED END LOCK
NO. 18 GAUGE AND LIGHTER

NO. 17 - INSIDE LOCK
WOOD TANK LINING
NO. 18 GAUGE AND LIGHTER

NO. 18 - LOCK SEAM
NO. 18 GAUGE
AND LIGHTER

NO. 19
(A) CAST BRASS FITTINGS   (B) WROUGHT COPPER FITTINGS   (C) LEAD JOINTS

Fig. 7-1 (*continued*)

until the solder drops disappear, and continue to apply solder until the joint is filled.

- Use a flux while soldering, applying it by dipping the solder into the flux from time to time.

- Do not touch or disturb the soldered joint until it has cooled below the melting point and has fully set. After the temperature has dropped below the melting point, wet cloths can be used to decrease the time normally required to cool the joint to room temperature. Note: A well-soldered joint should present a smooth, bright appearance, like polished silver. Wiping the joint before it cools destroys this appearance, and also is liable to produce roughness.

- The stronger the flux, the more corrosive it is, so be certain to clean off all traces of flux from the completed joint. This is best done by using water.

- If a smooth job is desired, finish the joint by removing the high spots with a medium file.

Note the following points about low-tin solders

- Careful cleaning is particularly necessary with the low-tin solders.

- The low-tin solders alloy more slowly, so stronger flux solutions are necessary.

- Tight joints are particularly important with these solders.

- The low-tin solders require higher metal temperatures than the more commonly used 50-50 or 60-40 tin-lead solders.

If the joint has a small enough diameter to be heated all over (as in water service tubing, for example), there should be no trouble with low-tin solders if the foregoing procedure is used. On large-diameter joints, however, pretinning both sides of the joint, after proper cleaning and fluxing, is an additional help.

The same basic soldering procedure applies to the high-lead silver solders, except that a temperature about 250°F higher is required to make the solder flow. The high-lead silver solders can be used where higher application temperatures are not a handicap. However, because of the higher temperature required, the flux used for 50-50 solder cannot be used with the high-lead silver solders. Special fluxes for this purpose are available, and flux manufacturers should be consulted.

INSPECTING SOLDERED JOINTS

The inspection of a soldered joint is generally a visual process, whether on the production line or elsewhere. As a matter of fact, visual inspections are most likely to occur on the production line, largely because there is insufficient time to apply electrical or mechanical tests.

The types of soldered joints that fail to meet visual inspection standards are:

- Joints with excess solder
- Joints with no solder at all
- Joints with insufficient solder
- Joints containing flux residue
- A joint in which solder functions as a ground or causes a short
- A joint with points protruding from the solder
- Joints that appear to have been made with a poor flow of solder
- Joints that appear to have been moved before being completely cooled

Movement of the soldered joint before it has completely cooled will cause it to be materially weakened. Premature movement all too often causes the solder to fracture, which in turn may result in joint failure.

The amount of solder applied to a joint determines why most joints are rejected during inspection. For example, some operators apply entirely too much solder to the joint. This can be corrected before final inspection by removing the excess solder. Try dipping the scrap end of unsoldered standard conductor into flux and pressing it between the terminal and the hot iron tip. This usually causes the excess solder to be drawn off into the strand.

Joints that have insufficient solder are weak and will usually fail if subjected to heavy vibrations. Such joints (and those to which no solder has been applied) are regarded as not being repairable, and they are rejected.

Sometimes excess flux will remain in the joint after soldering. If the soldered parts (wire, tube, sheet, etc.) are loose, the joint is defective and must be rejected. If no movement is possible, however, the flux can be removed by reheating the joint.

If the soldered joint has resulted in the shorting or grounding of electrical equipment, it should be rejected. This also applies to the formation of points on the surface of the solder, for these can disrupt a high-voltage circuit. Sometimes the solder does not have the opportunity to flow evenly through the joint, and reheating will correct this defect prior to the final inspection.

TORCH SOLDERING PROCEDURE

An oxyacetylene torch is generally not recommended for soldering because the temperature of the flame is too high. It usually destroys the protective action of the flux. This can be avoided by directing the full force of the flame away from the spot being soldered. This form of indirect heating can also be used with other types of gas flame soldering when the temperatures become too high for the metal or metal alloy.

The air-acetylene torch flame should show a bright, sharply defined inner cone and a pale blue outer flame. A yellow flame indicates that the acetylene pressure is inadequate, that the needle valve is not opened sufficiently, or that the soldering tip is clogged. The temperatures reached with this form of gas flame soldering are somewhat lower than those of the oxyacetylene flame.

Still lower flame temperatures can be obtained by using propane, natural gas, or butane. oxygen burned with fuel gases will produce higher temperatures than air burned with the same gases. The former will also produce a sharper, more well-defined flame than a torch operating on the Bunsen burner principle. The Bunsen burner units are generally gasoline-filled torches, and they produce a widespread flame that is somewhat bushy in appearance.

8. IDENTIFYING METALS

The first step in welding is to identify the metal. Correctly identifying the base metal provides the welder with the necessary information to choose an appropriate welding process. It also allows the welder to decide on the type of weld to use, the type of weld preparation, the composition of the filler rod (if used), and other factors that contribute to a successful weld. The inability to correctly identify the metal greatly increases the possibility of a poor weld.

The easiest and most accurate method of identifying a metal would be to send it to a laboratory for chemical analysis and/or machine testing. Unfortunately, this would be both expensive and time consuming, factors that make this method highly impractical for small welding shops and the individual welder working alone.

The metal identification methods described in this chapter are not comprehensive ones, but they can be useful in narrowing down the identification. For example, ferrous metals can be identified by the spark test, whereas nonferrous metals do not produce spark streams.

IDENTIFY BY USE

Metals are often easily identified by their use. Engine blocks, gears, shafts, and so forth can be expected to be constructed from certain types of metal (cast iron, forged steel, etc.). The manufacturer may specify this information in the literature accompanying the metal component or part, or it may be gained by contacting the manufacturer directly.

IDENTIFY BY SPECIAL MARKS OR PAPERWORK

Some metals are identified by codes or symbols stamped on their surface by the supplier. Examples are nickel alloys, such as Monel and Inconel, and wrought iron. The metal may also be identified in the drawing requirements (bill of materials) for a particular application.

Whenever possible, try to obtain a *material certificate*. It should include the manufacturer's name and the commercial brand name, heat number, dimensions, chemical analysis (composition), mechanical properties, delivery hardness, and finish of the metal item.

IDENTIFY BY SURFACE CONDITION

Sometimes it is possible to identify a metal by how the surface feels to the touch. For example, copper, brass, bronze, nickel, Monel, Inconel, lead, and wrought iron are very smooth. On the other hand, stainless steel is slightly rough to the touch, and both the low-alloy and high-carbon steels often have forging marks on the surface that make them seem a bit rough.

IDENTIFY BY FINISHED AND UNFINISHED COLOR

The finished (machined) and unfinished color of a metal can be used as a means of identification. Table 8-1 indicates color identifications for a number of commonly used metals and metal alloys. Colors are specified for both unfinished, non-fractured metal surfaces and finished (machined) surfaces. This means of identification leaves much to be desired insofar as the various steels are concerned (or in the case of Monel metal and nickel, for that matter).

IDENTIFY BY FIELD TESTS

Welders working alone in the field or in small shops, where special testing equipment is unavailable, must rely on other methods to identify the metal. They will be required to rely on their own experience plus some useful field tests that are not as accurate as chemical analysis or machine testing, but nevertheless are very useful and surprisingly accurate. These tests include: (1) the fracture test, (2) the spark test, (3) the flame test, (4) the chip test, (5) the magnetic test, and (6) the file scratch test.

Table 8-1   Colors of Finished and Unfinished Surfaces of Selected Metals

Metal Type	Unfinished Nonfractured Surface Color	Finished (Machined) Surface Color
Alloy steel	Dark gray	Bright gray
Aluminum and aluminum alloys	Very light gray	Very white
Beryllium	Steel gray	—
Brass	Green, brown, or yellow	Red to whitish-yellow.
Bronze	Green, brown, or yellow	Red to whitish-yellow.
Cast steel	Dark gray	Bright gray
Chromium	Steel gray	—
Columbium	Yellowish-white	—
Copper	Reddish-brown to green	Bright copper red
Gray cast iron	Dark gray	Light gray
High-carbon steel	Dark gray	Bright gray
Inconel	White	—
Lead	White to gray	White
Low-alloy steel	Blue gray	Bright gray
Low-carbon steel	Dark gray	Bright gray
Magnesium	Silver-white	—
Malleable iron	Dull gray	Light gray
Manganese steel (under 2% manganese)	Gray-white (resembles color of iron)	—
Manganese steel (14%)	Dull	—

(continued)

Table 8-1 *(continued)*

Metal Type	Unfinished Nonfractured Surface Color	Finished (Machined) Surface Color
Medium-carbon steel	Dark gray	Bright gray
Molybdenum	Silvery-white	—
Monel metal	Dark gray, smooth	Light gray
Nickel	Dark gray	White
Tantalum	Gray	—
Tin	Silver-white	—
Titanium	Steel gray	—
Tungsten	Steel gray	—
Vanadium	Bright white	—
White cast iron	Dull gray	Seldom machined
Wrought iron	Light gray	Light gray
Zinc	Dark grey	—
Zirconium	Silver-white	—

Fracture Test
Many metals can be identified by the color of their fractured surface when a part is broken, or by studying a chip after it has been removed from the surface with a chisel and hammer (see Table 8-2).

Spark Test
A method used to identify ferrous metals is the spark test (see Table 8-3). This is neither a difficult nor an expensive method of testing, but it does require some knowledge of different spark stream characteristics.

The spark test is administered by lightly touching the surface of the metal to a spinning, power-driven grinding wheel

Table 8-2 Fracture Test

Metal Type	Comments
Aluminum and aluminum alloys	White
Brass	Not used
Bronze	Not used
Cast steel	Bright gray
Copper alloys	Not used
Deoxidized copper	Red
Gray cast iron	Brittle fracture; dark gray fractured surface
High-carbon steel	Very light gray
High-manganese steel (11-14% manganese)	Very coarse-grained
Inconel	Light gray
Lead	White, crystalline grain structure
Low-alloy steel	Medium gray
Low-carbon steel	Bright gray
Magnesium	Not used
Malleable cast iron	Brittle fracture surface; dark gray
Manganese steel (14%)	Coarse-grained fracture
Medium-carbon steel	Very light gray
Monel	Light gray
Nickel	Almost white to white
Nodular cast iron	Brittle fracture surface
Stainless steel (300 series)	Varies widely depending on type
Stainless steel (400 series)	Varies widely depending on type
Tin	Not used

(continued)

Table 8-2 *(continued)*

Metal Type	Comments
Titanium	Not used
Tungsten	Brittle grain
White cast iron	Brittle
Wrought iron	Bright gray, fibrous grain pattern

(emery wheel). The spark stream (pattern) is then observed against a black background. Spark streams will vary in color, volume, length, and size depending on the carbon content of the metal (see Figure 8-1 and Table 8-4).

NOTE

The length of the spark stream is determined more by the amount of pressure exerted against the grinding wheel than by the properties of the metal itself. Spark stream length should not be regarded as an important characteristic.

CAUTION

A spark test may generate sparks that can cause an explosion if they contact certain volatile fumes or substances. The sparks thrown off by the metal are not electrical sparks, are they?

Flame Test

Metals will turn a distinctive color when a torch flame is directed against the surface (see Table 8-5). Metals also melt at different rates, and their molten weld puddles and slag may differ.

NOTE

Do not use the flame test if you suspect the metal is magnesium, because the magnesium may ignite under a torch flame in the open atmosphere.

Table 8-3 Spark Stream Term

Term	Comments
Spark stream or pattern 	The spark created by the grinding wheel. Composed of the carrier lines and spark bursts.
Carrier lines or shafts 	The straight lines starting at the point of contact on the grinding wheel and extending outward at angles.
Spark bursts 	Small multi-branched clusters (*sprigs*) occurring normally along or at the end of a carrier line. Some carrier lines end in three-pronged *forks* instead of clusters.
Arrows 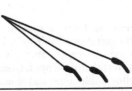	Some carrier lines end in continuous or broken arrow shapes.

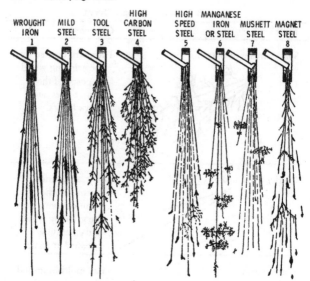

Fig. 8-1 The spark test for various types of steels.

Chip Test

A cold chisel and hammer can be used to cut the metal for the chip test. The metal is identified by how easy or difficult it is to cut the surface, whether the cut edges are smooth or rough, and whether the cut is continuous or results in brittle, small chips (see Table 8-6).

Magnetic Test

The magnetic test is performed with a small magnet placed against the surface of the metal (see Table 8-7). If the magnet fails to attach itself to the surface, the metal is clearly non-magnetic. Nonmagnetic metals are the nonferrous types, such as copper and copper alloys, aluminum and aluminum alloys,

Table 8-4   Spark Test for Selected Metals

Metal Type	Comments
Alloy steels	Spark stream similar to plain carbon steel except for changes proportional to the amount of alloy or alloys. Color changes from straw to deep orange, or to red. The stream pattern length shortens, with carrier lines appearing heavier. Carbon spark bursts are reduced in number.
Aluminum and aluminum alloys	Not used.
Brass	Not used.
Bronze	Not used.
Chromium	The particles seem to follow a broken line with a very slight explosion. Just before they disappear, the color is chrome yellow and shows no trace of a carbon spark.
Copper	Not used.
Copper alloys	Not used.
Gray cast iron	Produces a small-volume stream. Red carrier lines with sprigs or spark bursts similar to those of low-carbon steel.
High-carbon steel (carbon content of .45% or more)	Yellow carrier lines with numerous bright spark bursts. The iron lines of the spark stream are practically eliminated. Star-like explosions that often divide and subdivide are increased, causing a beautiful display of sparks.

(continued)

Table 8-4 (continued)

Metal Type	Comments
High-manganese steels (11–14% manganese)	Not used.
High-sulphur steel	Carrier lines swell. Sprigs and spark bursts not very similar and similar to low carbon steel.
Inconel	Not used.
Lead	Not used.
Low-alloy steel	The spark stream is characterized by a division, or forking, of the luminous streak.
Low-carbon steel (Carbon content of .20% or less)	Long, yellow carrier lines with little noticeable spark bursts.
Malleable cast iron (annealed)	Moderate stream volume. Stream color red, changing to yellow at end. Many fine, repeating spurts.
Manganese steels (under 2% manganese)	Large, bright white spark bursts spread out in a fan shape.
Manganese tool steel	Differs widely from the carbon spark by shooting or exploding at right angles from its line of force. Each dart of the manganese spark is subdivided into a number of white globules
Medium-carbon steel (.20% to .45% carbon)	Noticeable sprigs or spark bursts forming at ends of yellow lines.
Monel	Not used.
Nickel	Very small stream volume, orange in color. No spurts.

(continued)

Table 8-4 (continued)

Metal Type	Comments
Nodular cast iron	Spark stream has lighter to orange section near the grinding wheel. Lower section is longer than in the case of gray cast iron, with fewer spark bursts and a near-yellow color.
Stainless steel (300 series)	Varies depending on alloy content.
Stainless steel (400 series)	Varies depending on alloy content.
Tin	Not used.
Tool steel (lower grades: 50–100 point carbon content	The spark stream is characterized by iron lines that become less and less conspicuous. The forking of the luminous streak occurs very much more frequently, often subdividing. The lower the carbon content, the fewer the sparks, and the farther these sparks occur from the source of heat.
Tungsten	Not used.
Tungsten-chromium die steel	Small stream volume. Red stream with many fine, repeating spurts.
White cast iron	Very small stream volume. Stream color red, changing to yellow at end. Many fine, repeating spurts.
Wrought iron	Wrought iron creates a spark stream consisting of small particles flowing away from the point of contact in a straight line. The stream becomes broader and more luminous some distance from the source of heat, and then the particles disappear as they started.
Zinc	Not used.

Table 8-5 Flame Test for Selected Metals

Metal Type	Comments
Alloy steel	Impossible to establish uniform flame characteristics because of the widely varying composition of the alloy steels
Aluminum and aluminum alloys	Melts without any color
Brass	Not used
Bronze	Not used
Cast iron	Melts slowly
Copper	Not used
Copper alloys	Not used
Gray cast iron	
High-carbon steel	Turns a bright red just before melting and then melts quickly
High-manganese steels (11–14% manganese)	Shows color
High-sulphur steel	Turns a bright red just before melting and then melts quickly
Iconel	Not used
Lead	Melts fast
Low-alloy steel	Turns a bright red just before melting and then melts quickly
Low-carbon steel	Turns a bright red just before melting and then melts quickly
Magnesium	Ignites when it comes in contact with air and a flame

(continued)

Table 8-5 (continued)

Metal Type	Comments
Medium-carbon steel	Turns a bright red just before melting and then melts quickly
Monel	Not used
Nickel	Not used
Nickel alloys	Not used
Stainless steel (300 series)	Melts fast
Stainless steel (400 series)	Melts fast
Tin	Melts fast
Titanium	Not used
White cast ion	Turns a dull red before melting and then melts slowly
Wrought iron	Turns a bright red just before melting and then melts quickly
Zinc	Melts fast

Table 8-6 Chip Test for Selected Metals

Metal Type	Comments
Alloy steel	Impossible to establish uniform chip characteristics because of the widely varying composition of the alloy steels.
Aluminum and aluminum alloys	Chips easily. Very smooth chips. Saw edges where cut. Chip can be as long and continuous as desired.
Brass	Chips easily. Very smooth chips. Saw edges where cut. Chip can be as long and continuous as desired.

(continued)

Table 8-6 *(continued)*

Metal Type	Comments
Bronze	Chips easily. Very smooth chips. Saw edges where cut. Chip can be as long and continuous as desired.
Copper	Chips easily. Very smooth chips. Saw edges where cut. Chip size can be as long and continuous as desired.
Gray cast iron	Difficult to chip. Chips break off from base metal in small fragments.
High-carbon steel	Very hard metal and difficult to chip. Chips can be any size or shape, continuous and as long as desired. Edges are lighter in color than those of low-carbon steels.
High-sulphur steel	Chips easily. Chisel cuts a continuous chip with smooth edges.
Lead	Chips very easily. Chips can be any size or shape, and as long as desired.
Low-alloy steel Low-carbon steel	Chips easily, with smooth edges. Chisel can cut a continuous chip if desired.
Malleable cast iron	Tougher metal than other cast irons. Difficult to chip. Chips do not break short.
Medium-carbon steel	Chips easily. Chisel cuts a continuous chip with smooth edges.
Molybdenum Monel metal	Chips easily. Chips have smooth edges and can be as long as desired.
Nickel	Chips easily. Chips have smooth edges and can be as long as desired.
Nickel alloys	Chips easily. Chips have smooth edges and can be as long as desired.

(continued)

Table 8-6 (continued)

Metal Type	Comments
White cast ion	Chipping results in small, broken fragments. Brittleness of the metal prevents formation of a chip path with smooth edges.
Wrought iron	Chips easily. Very smooth edges on cut side of chip. Chip size can be as long and continuous as desired.

Table 8-7 Magnetic Test for Selected Metals

Metal Type	Comments
Alloy steel	Reduced magnetism, depending on the alloying element.
Aluminum and aluminum alloys	Nonmagnetic.
Brass	Nonmagnetic.
Bronze	Nonmagnetic.
Cast steel	Magnetic.
Copper	Nonmagnetic.
Gray cast iron	Strongly magnetic.
High-carbon steel	Strongly magnetic.
High-sulphur steel	Strongly magnetic.
Inconel	Nonmagnetic.
Lead	Nonmagnetic.
Low-carbon steel	Strongly magnetic.
Magnesium	Nonmagnetic.
Malleable cast iron	Strongly magnetic.
Manganese steels (under 2% manganese)	Magnetic.
Medium-carbon steel	Strongly magnetic.

(continued)

Table 8-7 (continued)

Metal Type	Comments
Monel metal	Slightly magnetic.
Nickel	Magnetic.
Nickel alloys	Some are magnetic, some are weakly magnetic, and others are nonmagnetic. Their magnetism can seldom be determined from analysis.
Stainless steel (300 series)	Nonmagnetic.
Stainless steel (400 series)	Magnetic.
Tin	Nonmagnetic.
Titanium	Nonmagnetic.
White cast iron	Strongly magnetic.
Wrought iron	Strongly magnetic.
Zinc	Nonmagnetic.

Table 8-8 Brinell Values Compared to the File Scratch Test

Brinell Hardness Values	File Action	Probable Metal Type
100 BHN	File bites into the surface very easily.	Low-carbon steel
200 BHN	File removes metal with slightly more pressure.	Medium-carbon steel
300 BHN	Metal exhibits its first real resistance to the file.	High-carbon steel or high-alloy steel

(continued)

Table 8-8 *(continued)*

Brinell Hardness Values	File Action	Probable Metal Type
400 BHN	File removes metal with difficulty.	Unhardened tool steel
500 BHN	File marks the surface and just barely removes metal.	Hardened tool steel
600 BHN	File slides over the surface without removing metal; file teeth are dulled.	Metal harder than the file

Courtesy James F. Lincoln Arc Welding Foundation

and magnesium. Ferrous metals, such as carbon steels, low-alloy steels, pure nickel, and martensitic stainless steels are strongly magnetic, and the magnet will attach itself securely to the metal surface. Some metals, however, are only slightly magnetic, but experience will eventually enable the welder to distinguish between strongly and slightly magnetic metals.

File Scratch Test

A simple scratch test with a sharp mill file is sometimes used to make a rough determination of metal hardness in order to identify different types of steels. The file is drawn across the metal surface and the scratch is compared to selected Brinell values to determine the hardness of the metal (see Table 8-8).

9. CAST IRON

Cast iron is a mixture of 91 to 94 percent iron and varying proportions of other elements, the most important being carbon, silicon, manganese, sulphur, and phosphorous. The term *cast iron* refers to a class of metals sharing most of the same characteristics. The four basic types of cast iron are: (1) gray cast iron, (2) white cast iron, (3) malleable cast iron, and (4) nodular cast iron (see Table 9-1).

Cast-iron welding is used most commonly to correct casting defects before the castings are placed in service, to repair worn or broken castings already in service, or to fabricate assemblies. Welding machine support frames and bases, housings, enclosure assemblies, and pipe fittings are typical applications.

ALLOY CAST IRON

Nickel, molybdenum, chromium, copper, aluminum, and other elements can be added to gray cast iron to produce an alloy possessing certain specifically desired characteristics (greater corrosion resistance, higher strength). For example, a nickel cast iron will have greater corrosion and wear resistance than gray cast iron, and a molybdenum cast iron will posses higher tensile strengths.

NOTE

The welder should take the characteristics of the alloying element into consideration and proceed accordingly when welding alloy cast irons. Aluminum cast irons, for example, present the problem of aluminum oxides forming on the surface during welding.

GRAY CAST IRON

Gray cast iron (also sometimes referred to as *gray iron* or *pig iron*) is an alloy composed of iron, carbon, and silicon. Traces of phosphorous are also usually present. The chemical

Table 9-1 A Summary of Welding Procedures for Different Types of Cast Iron

Cast Iron Type	Procedure	Treatment	Properties
Gray iron	Weld with cast iron	Preheat and cool slowly	Same as original
Gray iron	Braze weld	Preheat and cool slowly	Weld better; heat affected zone as good as original
Gray iron	Braze weld	No preheat	Weld better; parent metal hardened
Gray iron	Weld with steel	Preheat if at all possible	Weld better; parent metal may be too hard to machine; if not preheated, needs to be welded intermittently to avoid cracking
Gray iron	Weld with steel around studs in joint	No preheat	Joint as strong as original
Gray iron	Weld with nickel	Preheat preferred	Joint as strong as original; thin hardened zone; machinable

(continued)

Table 9-1 *(continued)*

Cast Iron Type	Procedure	Treatment	Properties
Malleable iron	Weld with cast iron	Preheat, and post-heat to repeat malleablizing treatment	Good weld, but slow and costly
Malleable iron	Weld with bronze	Preheat	As strong, but heat-affected zone not as ductile as original
White cast iron	Welding hot recommended		
Nodular iron	Weld with nickel	Preheat preferred; postheat preferred	Joint strong and ductile, but some loss of original properties; machineable; all qualities lower in absence of preheat and/or postheat

Courtesy The James F. Lincoln Arc Welding Foundation

analysis of a gray cast iron will vary in accordance with its composition. For example, the silicon content may vary from 1.0 to 3.0 percent and the carbon content from 2.5 to 4.0 percent (in the free or graphitic state). The amounts will vary according to the use for which the gray cast iron is produced. Gray cast iron contains approximately 3.0 to 3.7 percent carbon. The tensile strength of gray cast iron ranges from 10,000 to 60,000 psi, depending on the particular class.

NOTE

The American Society for Testing and Materials (ASTM) has devised a classification system based on the tensile strengths of different gray cast irons. A gray cast iron having a minimum tensile strength of 10,000 psi belongs to class 10; one with a minimum tensile strength of 20,000 psi belongs to class 20; and so forth. There are seven classes altogether.

The melting point of gray cast iron is approximately 2150°F. Its ductility is rather low, which is indicated by limited distortion at breaks, and its impact and shock resistance is almost nonexistent. Gray cast irons generally exhibit ease of machining and weldability. They are also easily cast in a wide variety of forms. Note: Gray cast iron is not malleable at any temperature.

MALLEABLE CAST IRON

Malleable cast iron is produced by annealing or heating white cast iron over a prolonged period of time and then allowing it to cool slowly. This heat treatment results in a greater resistance to impact and shock, higher strength, and greater ductility. Consequently, malleable cast iron is better able to withstand these types of strain than gray cast iron.

The tensile strength of malleable cast iron ranges from 30,000 to 100,000 psi. The strength depends upon whether the normal base structure is ferrite or pearlite. The lower tensile strengths (53,000 psi or below) belong to the ferrite malleable cast irons; the higher ones (60,000–100,000 psi) to the pearlite group. Most malleable cast iron is produced with a tensile strength of 40,000–50,000 psi. It should be noted that as the tensile strength of a malleable cast iron increases, the ductility decreases.

The carbon content of malleable cast iron is approximately 2 to 3 percent and must be in the combined form (free or graphitic carbon cannot be made malleable). Malleable cast iron is stronger and tougher than gray cast iron. It possesses good machining characteristics, but because of its malleability it has limited weldability.

Recommended joining processes for working with malleable cast iron include oxyacetylene welding and braze welding, with the latter preferred because of its lower temperatures. High welding temperatures will cause the malleability characteristic to break down and a reversion to the characteristics of white cast iron. For this reason, welding processes that generate a high heat (shielded metal arc, carbon arc, etc.) are not recommended for working with malleable cast iron *unless* temperatures of approximately 1400° to 1450°F or below are used.

NODULAR CAST IRON

Nodular cast iron (also referred to as *ductile cast iron* or *spheroidal cast iron*) is so named because the graphite present in its composition takes the form of nodules, rather than flakes. This gives nodular cast iron greater shock resistance than gray cast iron. Graphite nodules are formed by adding magnesium to the molten metal during the production process.

The melting point of nodular cast iron is slightly below that of gray cast iron. Special heat treatment and the addition of alloying elements results in a metal that combines the more desirable characteristics of both gray cast iron and steel. The principal element added to obtain these characteristics is magnesium.

The tensile strengths of nodular cast iron will range from 60,000 to 120,000 psi, depending on the composition of the base metal, the process used to obtain the nodule structure, and other factors.

NOTE

Both oxyacetylene welding and oxyacetylene braze welding can be used to weld nodular cast iron.

WHITE CAST IRON

White cast iron, which is also referred to as *high-strength cast iron*, is produced from pig iron by causing the casting to cool very rapidly. The rate of cooling is too rapid to allow the carbon to separate from the iron carbide compound. Consequently, the carbon found in white cast iron exists in the combined form.

White cast iron is very hard and brittle, and does not lend itself well to machining. As a result, this type of cast iron is not often used in its original state for castings. Because of its hardness, it is frequently used as a wear-resistant outer surface for inner cores of gray cast iron. These white cast iron surfaces are produced by causing the molten metal to flow against heavy iron chills placed in the mold. This causes rapid cooling that results in a very hard surface. Castings produced in this manner are referred to as *chilled-iron castings*. One of the major uses of white cast iron is in the production of malleable iron castings.

Generally, white cast iron will have a tensile strength of 40,000–50,000 psi or more. The melting point of white cast iron is 2300°F, or slightly higher than that of gray cast iron.

WELDING AND JOINING CAST IRON

Cast iron is less weldable than low-carbon steel because it contains greater amounts of carbon and silicon. As a result, it tends to be more brittle than low-carbon steel and has a tendency to crack in the heat-affected zone unless proper preheating, post-cooling, and low welding temperatures are used.

NOTE

The *heat-affected zone* (often abbreviated HAZ) is the area of the base metal along the joint seam that is not melted, but whose structure has been altered by the welding or cutting heat.

Proper edge or area preparation is very important when welding cast iron. All surface defects must be completely removed. Sufficient space must be provided in the joint for satisfactory filler metal deposition with the minimum penetration allowed. To minimize cracking, a ductile material should be chosen as the filler metal.

It is common practice to preheat the castings and to provide protected slow cooling after welding. The slow cooling is done to reduce residual stresses and avoid cracking. Preheating slows the cooling rate, permitting the formation of less brittle structures. It also permits the whole casting to contract together with the weld material, thereby reducing residual stresses.

Welds in cast iron, if of sufficient thickness, may be strengthened by the mechanical method of studding. Steel studs approximately 1/4 to 3/8 inch in diameter should be used. The cast iron should be beveled to form a vee, drilled, and tapped along the vee, so that the studs may be screwed into the casting. The studs should project about 3/16 to 1/4 inch above the cast-iron surface. They should be long enough to be screwed into the casting to a depth of at least the diameter of the studs (see Figure 9-1).

Oxyacetylene Welding

The oxyacetylene welding process can be used to weld both gray and nodular cast irons in all thicknesses. It is not recommended for welding malleable cast iron because it tends to promote brittleness.

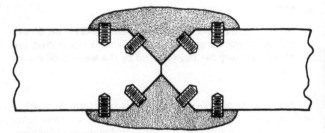

Fig. 9-1 Stud reinforcement of cast iron.

Note the following:

- The proper use of the welding rod and the flux (if the situation requires) is very important. The end of a length of 1/4-inch cast-iron welding rod should be introduced into the outer cone of the flame, heated, dipped into the flux, and then placed with the flux end in the molten puddle. The heat of the molten metal on which the welding flame continues to play will melt the rod gradually, and the surface of the puddle will gradually rise with the addition of this metal.

- The rod should never be held above the weld and melted drop by drop into the puddle. Also, be careful to fuse the side of the v-groove ahead of the advancing puddle so that the molten metal is never forced onto colder metal. If that happens, it will cause an adhesion with little or no strength at that spot.

- When gas bubbles or white spots appear in the puddle or at the edges, flux should be added and the flame played around the spot until the impurities float to the top. These impurities are skimmed from the weld with the hot welding rod, to which they adhere. Tapping the hot rod against the table eliminates them entirely. This removal of dirt must be done carefully and systematically, for impurities left in the weld constitute defects and result in a weak joint.

- The rod should be added to the molten metal until that section of the vee is built up slightly above the level of the rest of the piece. When one section an inch or so long is built up, the bottom of the vee adjacent to it is melted as the operation is repeated. Of course, care must be taken to keep both the end of the built-up section and the sides of the vee in complete fusion with the puddle.

- Cast-iron welding should be done as quickly as possible. When finished, cover the completed weld with a non-flammable insulating material or bury it in an annealing bin to cool slowly.

- If preheating is unnecessary or inadvisable, care should be taken at the point of welding not to heat the casting too long or too much at one time. The recommended procedure is to apply the welding heat as briefly as possible, and then to allow the casting to cool for a somewhat longer period. Repeat this procedure until the weld is completed.

- Oxyacetylene welding requires a higher preheat process than shielded metal arc welding.

Shielded Metal Arc Welding

Shielded metal arc welding (SMAW) is commonly used to weld large castings of gray cast iron. Either machinable or nonmachinable electrodes can be used.

There are two basic types of machinable electrodes: (1) the 100-percent nickel core electrode and (2) the nickel-iron base electrode (see Table 9-2). Both types produce a relatively soft, ductile weld deposit that can be machined after cooling. These electrodes are commonly used to repair broken or cracked castings, fill surface voids, or weld dissimilar metals—cast iron to steel, for example.

Table 9-2 Machinable Electrodes for Welding Cast Iron

AWS Classification	Comments
ENi-Ci	<ul><li>100-percent nickel core</li><li>General-purpose electrode used with DC electrode positive (DCEP) or AC current</li><li>Used to weld thin and medium-thickness sections</li><li>Used to weld castings with low phosphorous content</li><li>Used to weld with little or no preheating</li></ul>
ENi-FeCi	<ul><li>Nickel-iron base core</li><li>Used with DC electrode positive (DCEP) or AC current to weld heavy cast-iron sections</li><li>Used to weld castings with high phosphorous content</li><li>Used to weld high-nickel alloy castings where high-strength welds are required</li><li>Recommended for welding nodular cast iron</li></ul>

A nonmachinable electrode, on the other hand, has a mild steel core. Because the electrode coating melts at low temperatures, a low welding current can be used to weld the cast iron. It produces a very hard and water-proof weld deposit that is not machined afterward. This electrode is used for repairing automotive engine blocks, compressor blocks, water jackets, pump parts, and other components that require a tough joint without the need for machining.

Note the following:

- Use a 1/8-inch diameter electrode to keep the heat down. The electrode is made positive, the work is negative, and the current is approximately 80 amps. This low current level is used to satisfy the heat conditions. The electrode itself will carry considerably more current, but the requirements of cast-iron welding make the use of higher heat inadvisable.

- Because of the low current used (80 amps on a 1/8-inch electrode), the hardening effect usually present along the line of fusion is materially reduced. As a result, the weld is more machinable than is usually the case when other electrodes are employed.

- The welding of gray cast iron should be done intermittently. In some cases, *skip welding* is used to make a weld not longer than 8 inches at one time. Immediately after each bead is deposited, it should be lightly peened, thoroughly cleaned, and allowed to cool before the next bead is applied. Care should be taken to keep the work clean and not allow it to become too hot. A good rule in reference to cast iron is to keep the work clean and cold.

Gas Tungsten Arc Welding

Gas tungsten arc welding (GTAW), or TIG welding, can be used to weld gray cast iron, but it does not provide any specific advantages over less expensive welding processes. A higher preheat is usually recommended. Filler materials in the form of rods that have a chemical composition similar to those used for SMAW can be used.

Gas Metal Arc Welding

Gas metal arc welding (GMAW), or MIG welding, is recommended for welding ductile and malleable cast iron, although it can be used for welding gray cast iron when productivity is important. GMAW produces high deposition rates with limited weld penetration. It is recommended for welding thicknesses of 1/4 inch or thicker.

Flux Cored Arc Welding

Flux cored arc welding (FCAW) is a wire-feed welding process in which a continuous, consumable electrode wire is fed into the molten weld pool. FCAW produces high deposition rates with limited weld penetration. It is recommended for welding thicknesses of 1/4 inch or thicker.

Braze Welding

One very common use of braze welding is in the repair of gray iron castings. A copper-base alloy is generally used to either join fractured parts or to build up worn or missing sections.

Braze welding with copper alloy filler metals is actually very effective on any type of cast iron. The low temperature requirements of braze welding make the process particularly suitable for joining malleable iron.

Among the problems associated with using the braze welding process to weld cast iron are:

- The color of the copper alloy filler metal does not match that of the iron.

- The corrosion resistance of the weld metal differs from that of the base metal.

- Galvanic corrosion may become a problem after dissimilar metals are joined.

- The strength of a braze weld falls off rapidly as temperatures increase.

Preheating may be applied locally to cast iron, or it may extend over a wide area. It all depends on the size of the casting. Large castings require extensive preheating.

The preheating temperature is important. If it is too hot or too cold, the filler metal will not wet (tin) the joint. Cast iron seldom requires preheating to more than 1000°F, and lower temperatures are often used. The composition of the base metal will determine the correct preheating temperature. Note: The heat added during braze welding is normally of short duration.

Note the following:

- Clean the surfaces thoroughly, because the low braze welding temperatures do not remove oxides and other impurities. Note: These same low temperatures result in less distortion and fewer weld cracks. Failure of welds along the joint seam is usually caused by improper cleaning, fluxing, or tinning.

- Preheating is usually not necessary when braze welding, but in some cases a slight preheating may improve weld results. Note: Wetting and tinning may be impaired or made more difficult if the preheating temperature is either too low or too high.

- Braze welding cast iron does not require post-heating. Cool the weld slowly under suitable insulating materials.

- A BCuZn-A brazing filler metal is recommended for brazing welding cast iron.

- Use a neutral oxyacetylene flame for braze welding gray cast iron, and a slightly oxidizing flame for malleable cast iron.

- A suitable flux is required to keep the surface clean and assist capillary attraction during braze welding.

Brazing

Torch brazing of cast iron is used in limited production and/or repair and maintenance applications. It is used to join cast iron by heating the metal to a temperature above 800°F. The filler metal then flows between the closely fitted adjoining surfaces by capillary attraction. The type of brazing used to join cast iron is called *silver alloy brazing* because it uses an alloy of silver, copper, and/or zinc mixed with small amounts of other alloying elements.

Note the following:

- Fit the parts together with a clearance no greater than 0.002 to 0.006 inch. A wider separation will weaken the joint.

- Thoroughly clean the surface by removing any grease, oil, oxides, or other contaminants. Use sandpaper or emery paper for a mechanical cleaning. Pickling in acid is a common chemical cleaning method.

- Surface contaminants are also commonly removed from gray cast iron by suspending the parts in an open container of catalyzed molten salts and passing an electrical current through them.

- Use an appropriate flux to further clean the surface and prevent oxidation during brazing.

- A BCuZn-A brazing filler metal is recommended for brazing welding cast iron.

Soldering

Torch soldering is used to join cast iron by heating the metal to a temperature below 800°F. Most soldering of cast iron is done to repair castings or to fill surface imperfections. Although cast irons are generally difficult to solder, soldering has been used to join gray, malleable, and nodular cast iron ranging in thickness from 1/8 to 3/4 inch.

NOTE

Gray cast iron is particularly difficult to solder because its graphitic iron content prevents the molten solder from bonding properly. White cast iron is almost never soldered.

Note the following:

- Make certain the surface is absolutely clean before soldering, or surface contaminants will interfere with the wetting (tinning) process.

- File or machine the surfaces of castings to provide a tight fitup and to remove any surface impediments to wetting.

- Avoid overheating localized areas of the casting. Excessive heat may crack the casting.

- Use corrosive fluxes similar to those used for soldering stainless steel.

- Always cool the soldered joint uniformly to prevent the solder from breaking loose from the metal.

10. WROUGHT IRON

Wrought iron is produced by melting pig iron ingots in a puddling furnace. This process not only removes almost all of the carbon, but also other impurities such as silicon and manganese. The highly refined iron combines in a mechanical mixture with the iron oxide-silicate slag to produce the fibrous structure of wrought iron. Its good tensile strength, high ductility, high corrosion resistance, and fatigue resistance are all attributed to its structure.

The tensile strength of wrought iron is about 45,000 psi. Its melting point is approximately 2800°F, or several hundred degrees above the melting point of its own slag. Wrought iron is characterized by an extremely low carbon content. In fact, it has the lowest carbon content among the various commercial irons.

WROUGHT-IRON WELDING AND JOINING PROCESSES

Either oxyacetylene welding or braze welding can be used to weld wrought iron. Braze welded joints are not nearly as strong as those produced by oxyacetylene welding.

Oxyacetylene Welding

Oxyacetylene welding has traditionally been the welding process used to weld wrought iron. It should be noted, however, that these welds do not have a wrought-iron structure.

Note the following:

- Use a neutral oxyacetylene flame to weld wrought iron.
- Use either a low-alloy steel gas welding rod (RG 60) or a low-carbon steel gas welding rod (RG 45). Do not use high-carbon steel welding rods to increase weld strength. The results are not very satisfactory.

- No flux is required when welding wrought iron. The weld and adjacent surfaces have excellent protection against oxidation because the slag melts at a temperature considerably below that of the wrought iron base metal.

- Avoid excessive stirring of the molten weld puddle. Doing so often results in oxide formation caused by oxygen combining with elements in the weld puddle.

- Acceptable joint designs include the lap joints, T-joints, V-groove butt joints, fillet joints, and plug joints.

NOTE

Slag will sweat to the surface at a relatively low temperature when welding wrought iron with the oxyacetylene process. This is a false indicator of fusion. Keep the welding rod immersed in the molten puddle and continue to apply heat until the base metal breaks down and fusion is complete.

Braze Welding

In braze welding, the filler metal melts above 842°F but below the melting point of the base metal. In other words, no fusion of the base metal takes place as it does in oxyacetylene welding. Unlike in brazing or soldering, the filler metal is not distributed in the joint by capillary action. It is deposited in the joint, where it adheres to the base metal. Although the joint is not as strong as one formed by oxyacetylene welding, the low temperatures used in braze welding produce less stress and distortion.

Note the following:

- Use a neutral oxyacetylene flame to braze weld wrought iron.

- Use a bronze gas welding rod (RB CuZn-A), a nickel gas welding rod (RB CuZn-B), or a manganese bronze gas welding rod (R Cu Zn-C).

- Use a flux (either liquid, paste, or powder) designed for braze welding cast iron or wrought iron.

- Joint design is similar to that used in oxyacetylene welding.

I I. CARBON STEELS

It is not always easy to make a clear-cut distinction between iron and steel. This is particularly true if the carbon content of either one is close to the 2-percent level, because the carbon content is frequently used as a factor in classifying them. If the carbon content exceeds 2 percent, the metal is classified as one of the several irons (cast iron, wrought iron, etc.). On the other hand, a carbon content of less than 2 percent indicates that the metal should be classified as steel. Some common uses of carbon steel are listed in Table 11-1.

LOW-CARBON STEELS

Low-carbon steels (also referred to as *mild steels* or *mild-carbon steels*) are characterized by a carbon content not exceeding 0.30 percent. The minimum amount of carbon found in these steels ranges from 0.05 to 0.08 percent. A carbon content below these minimum levels (i.e., below 0.05 percent) indicates a type of pure iron referred to as *ingot iron*.

NOTE

Low-carbon steels, with less than 0.25 percent carbon and a sulfur and phosphorus content less than 0.04 percent, can be easily welded with most of the welding processes. There are no special precautions to be taken; the welding heat will not affect their basic properties.

The low-carbon steels have a melting point of approximately 2600 to 2700°F. The slag contained in these steels will melt at a lower temperature than the metal itself, making the use of a flux unnecessary. The low-carbon steels possess a number of excellent qualities that contribute to their being the most widely used steels in industry today. First, and probably foremost, they are relatively inexpensive. The low-carbon steels are also tough and ductile, and are generally easy to weld and machine.

Table 11-1   Carbon Steel Applications

Carbon Class	Carbon Range %	Typical Uses
Low	0.05–0.15	Chain, nails, pipe, rivets, screws, sheets for pressing and stamping, wire
	0.15–0.30	Bars, plates, structural shapes
Medium	0.30–0.45	Axles, connecting rods, shafting
High	0.45–0.60	Crankshafts, scraper blades
	0.60–0.75	Automobile springs, anvils, bandsaws, drop-hammer dies
Very high	0.75–0.90	Chisels, punches, sand tools
	0.90–1.00	Knives, shear blades, springs
	1.00–1.10	Milling cutters, dies, taps
	1.10–1.20	Lathe tools, woodworking tools
	1.20–1.30	Files, reamers
	1.30–1.40	Dies for wire drawing
	1.40–1.50	Metal-cutting saws

NOTE

Copper-bearing steels are low-carbon steels that contain approximately 0.20 percent copper. A copper content exceeding this amount generally causes the development of surface cracking in the base metal surrounding the weld. Copper-bearing steels are welded with the same procedures used for other low-carbon steels.

Note the following:

- Shielded Metal Arc Welding

    a. The shielded metal arc process is commonly used to weld mild steel in the low-carbon ranges. As the carbon content increases, it becomes necessary to apply preheat and post-heat treatment procedures in order to obtain a satisfactory weld.

b. Both mild steel electrodes (E60XX) and low-hydrogen electrodes (e.g., the E6015 or E6016) are recommended for use with low-carbon steels.

c. Weld cracking can be avoided when working with medium-carbon steels in the higher carbon ranges by employing special welding techniques and using suitable electrodes.

d. A post-heat treatment is recommended.

- Oxyacetylene Welding

    a. Commonly used to weld thin gauges of sheets or tubes.

    b. Because of the significant difference between the melting temperature of the mild steel and its slag, a flux is generally not used. The selection of a filler metal is important, however, and will depend on the analysis of the base metal. This, in turn, will determine the type of flame used. There are three ferrous welding rods recommended as filler metals for the low-carbon (mild) steels. They are: (1) An alloy-steel rod (AWS Classification GA 60); (2) A high-tensile steel rod (AWS Classification GA 60); and (3) A low-carbon (mild) steel rod (AWS Classification GA 50).

    c. The alloy-steel rod will produce a weld with a tensile strength in excess of 60,000 psi. The weld produced by the low-carbon steel rod will be somewhat lower, with a minimum tensile strength of 52,000 psi.

    d. A neutral flame or one with a slight excess of acetylene is recommended when using the alloy-steel rod. A slightly carburizing flame is also recommended for welding with either a high-tensile or a low-carbon steel rod (with the carburizing flame, a somewhat higher welding speed will be achieved). An oxidizing flame must be strictly avoided, because it will form oxides on the surface that may be included in the weld, thereby weakening it. Excessive sparking is one indicator of an oxidizing flame.

e. A backhand welding technique is preferred to a forehand one. In backhand welding, the flame is directed back against the weld, where it functions as a shield against contamination and thus reduces the possibility of oxide formation. Using this technique also results in less torch and rod manipulation, less agitation of the metal, and faster welding speeds.

f. Because of the significant difference between the melting temperature of the mild steel and its slag, a flux is generally not used.

g. The selection of a filler metal is important and will depend on the analysis of the base metal. This, in turn, will determine the type of flame used.

h. A post heat treatment is recommended

MEDIUM-CARBON STEELS

The carbon content of the *medium-carbon steels* ranges from 0.30 to 0.45 percent. These steels are not as easily welded as the low-carbon steels, and this is directly due to the increased carbon content. The medium-carbon steels are harder and stronger than the mild steels, but they tend toward weld brittleness in the high-carbon range.

Note the following:

- Low-hydrogen electrodes are recommended when using the SMAW process to weld medium-carbon steels.

- The E-6015, E-6016 and E-6018 electrodes require preheating the base metal to reduce weld cracking. These electrodes are used where greater amounts of weld deposit are required.

HIGH-CARBON STEEL

High-carbon steel contains 0.45 to 0.75 percent carbons. It finds common application in the manufacture of such items as knives, saws, bits, and tools. In other words, these manufactured items require hard, tough surfaces exhibiting a high resistance to wear. Some classifications of the carbon steels add a fourth subgroup with an even higher carbon content range (0.75 to 1.50 percent). These are referred to as *very-high-carbon steels*. Due to their extreme hardness and other characteristics that limit the weld quality, these carbon steels are seldom welded, except for repair purposes.

Note the following:

- The major problem with the high-carbon steels (and the very-high-carbon steels) is that they require controlled preheating and post-heating. The preheating and post-heating must be maintained within a specific temperature range. The following temperature ranges are recommended:

  a. Preheat steel with a carbon content of 0.45 to 0.65 percent to temperatures from 200 to 400°F (94 to 204°C).

  b. Preheat steel with a carbon content of more than 0.65 percent to temperatures from 752 to 1292°F (204 to 371°C).

  c. Post-heat all high-carbon steels to temperatures from 1098 to 1200°F (592 to 650°C).

- The adverse effects of the welding heat can be somewhat lessened by maintaining a fast welding speed.

- Low-hydrogen electrodes are recommended when welding high-carbon steels with the SMAW process. These electrodes will greatly reduce the possibility of cracking.

- Many high-carbon steels require post-heat treatment for stress relief.

12. ALLOY STEELS

Alloys of steel start with the steel itself. Steel is an alloy of iron and carbon. When steel is produced, certain other elements are present in small quantities:

- manganese
- phosphorous
- sulfur
- silicon

When these four foreign elements are present in their normal percentages, the product is referred to as *plain-carbon steel*. Its strength is primarily a function of its carbon content. The ductility of plain-carbon steel decreases as carbon content is increased. Its harden ability is quite low. Additionally, the properties of ordinary carbon steels are impaired by both high and low temperatures. They are subject to corrosion in most environments.

DIFFERENCES AMONG STEELS

The difference between plain-carbon and alloy steel is often debatable. Both contain carbon, manganese, and usually silicon. Copper and boron also are possible additions to both classes. Steels containing more than 1.65 percent manganese, 0.60 percent silicon, or 0.60 percent copper are designated as *alloy steels*. Steel is also considered to be an alloy steel if a definite or minimum amount of another alloying element is specified or required. The most common alloy elements are chromium, nickel, molybdenum, vanadium, tungsten, cobalt, boron, and copper, as well as manganese, silicon, phosphorous, and sulfur in greater amounts than are normally present. See Table 12-1.

Table 12-1 Alloying Elements for Steel

Element	Melting Point (°C)	Application	Result
Aluminum	658	Little aluminum remains in steel.	Deoxidizes and refines grain. Removes impurities.
Chromium	1615	Stainless steels, tools, and machine parts.	Improves hardness of the steel in small amounts.
Cobalt	1467	High-speed cutting tools.	Adds to cutting property of steel, especially at high temperatures.
Copper	1082	Sheet and plate materials.	Retards rust.
Lead	327	Machinery parts.	Lead and added tin form a rust resistant coating on steels.
Manganese	1245	Bucket teeth. Rails and switches.	Prevents hot shortness by combining with sulfur. Deoxidizes, increases toughness and abrasion-resistance.
Molybdenum	2535	Machinery parts and tools.	Increases ductility, strength, and shock resistance.
Nickel	1452	Stainless steels. Acid-resistant tools and machinery parts.	In large amounts—resists heat, adds strength, toughness, and stiffness to steel.

(continued)

244

Table 12-1 (continued)

Element	Melting Point (°C)	Application	Result
Phosphorus (provided by ore)	43	Some low-alloy steels.	Up to 0.05 percent increases yield strength.
Silicon	1420	Precision castings.	Removes the gases from steel. Adds strength.
Sulphur	120	Some machined pieces.	Adds to the steel's machinability.
Tin	232	Cans and pans.	Forms a coating on steel for corrosion-resistance.
Titanium	1800	Used in low-alloy steels.	Cleans and forms carbide.
Tungsten	3400	For magnets and high-speed cutting tools	Helps steel retain hardness and toughness at high temperatures.
Vanadium	1780	Springs, tools, and machine parts.	Helps to increase strength and ductility.
Zinc	420	Wire, palls, and roofing.	Forms a corrosion-resistant coating on steel.
Zirconium	1850	Machine parts and tools.	Deoxidizes, removing oxygen and nitrogen. Creates a fine grain.

Plain-Carbon Steel

Plain-carbon steels are classified into subgroups:

- low-carbon steels (less than 0.30 percent carbon)
- medium-carbon steels (between 0.30 and 0.80 percent carbon)
- high-carbon steels (more than 0.80 percent carbon)

Effects of Alloying Steels

Alloying elements are added to steel in small percentages—usually less than 5 percent. Alloying is used to improve strength or the ability to be hardened. With the addition of much larger amounts of alloying elements—often up to 20 percent—the alloy produced has special properties, such as corrosion resistance or stability at high or low temperatures Additions may be made during the steel-making process to remove dissolved oxygen from the melt. Note the melting point in the various metals and alloys in Table 12-2.

- Manganese, silicon, and aluminum are often used for deoxidation.
- Aluminum and—to a lesser extent—vanadium, columbium, and titanium are used to control austenitic grain size.
- Sulfur, lead, selenium, and tellurium are used for machinability.
- Manganese, silicon, nickel, and copper add strength by forming solid solutions in ferrite.
- Chromium, vanadium, molybdenum, tungsten, and other elements increase strength by forming dispersed second-phase carbides.
- Columbium, vanadium, and zirconium can be used for ferrite grain-size control.
- Nickel and copper are added to low-alloy steels to provide improved corrosion resistance.

Table 12-2 Melting Points of Metals and Alloys

Element	Melting Tempereture	Alloy
Carbon	3500	
	3400	
	3300	
	3200	
Chromium	3100	
Pure iron	3000	Wrought iron
Mild steel	2900	Stainless steel, 12%
	2800	chromium
	2700	Cobalt
Nickel	2600	Silicon
Stainless steel,	2500	
19% chromium	2400	
Manganese	2300	
	2200	Cast iron
	2100	
	2000	Copper
	1900	
Silver	1800	Brass
	1700	
	1600	Bronze
	1500	
	1400	
Aluminum	1300	Magnesium
	1200	
	1100	
	1000	Aluminum alloys
	900	Magnesium alloys
Zinc	800	
	700	
Lead	600	
	500	
Tin	400	

IDENTIFYING STEELS

Welders should be able to identify the metals they are welding. Besides testing the metal with a magnet to determine if it is ferromagnetic, the physical color of the metal is a basis for determining its composition. Verifying the melting point of the metal is a more accurate method of determining its composition. Most metals' melting points are available from a number of sources.

Identification Methods

Steels can be identified by a number of methods. Once the metal has been used and is about to be reused or welded for another purpose it may be necessary first to investigate the ability of the steel to do the job. Three of the ways to identify metals are shown in Tables 12-3 and 12-4, as well as in Figure 12-1. In the spark test, the metal sparks produced by a grinding wheel have distinct shapes and colors. After a bit of practice the welder is able to tell which steel is being tested.

HIGH-ALLOY CAST IRONS

For alloy cast irons that are not to be heat-treated, the alloy elements are often selected to alter the properties by affecting the formation of graphite or cementite (iron carbide, Fe_3C). This means modifying the morphology of the carbon-rich phase, or simply strengthening the matrix material. Alloy elements are often added in small amounts to improve strength properties or wear resistance. High-alloy cast irons are often designed to provide corrosion resistance, especially at high temperatures such as those in the chemical industry.

Austenitic gray cast irons are among the high-alloy cast irons, and they are quite common. They contain about 14 percent nickel, 5 percent copper, and 2.5 percent chromium. They possess good corrosion resistance to acids and alkalis at temperatures up to about 800°C (1400°F).

Table 12-3 Metal Identification by Appearance

	Alloy Steel	Copper	Brass and Bronze	Aluminum and Alloys	Monel Metal	Nickel	Lead
Fracture	Medium gray	Red color	Red to yellow	White	Light gray	Almost white	White crystalline
Unfinished surface	Dark gray: relatively rough; rolling or forging lines may be noticeable	Various degrees of redish brown to green due to oxides	Various shades of green, brown, or yellow due to oxides; smooth	Evidences of mold or rolls; very light gray	Smooth; dark gray	Smooth; dark gray	Smooth velvety; white to gray
Newly machined	Very smooth; bright gray	Bright copper red color dulls with time	Red through to whitish yellow; very smooth	very smooth; very light gray	Smooth very white	Very smooth; white	Very smooth; white

Table 12-3 (continued)

	White Cast Iron	Gray Cast Iron	Malleable Iron	Wrought Iron	Low-Carbon Steel and Cast Steel	High-Carbon Steel
Fracture	Very fine silvery white silky crystalline formation	Dark gray	Dark gray	Bright gray	Bright gray	Very light gray
Unfinished surface	Evidence of sand mold; dull gray	Evidence of sand mold; very dull gray	Evidence of sand mold; dull gray	Light gray; smooth	Dark gray; forging marks may be noticeable; cast-evidences of mold	Dark gray rolling or forging lines may be noticeable
Newly machined	Rarely machined	Fairly smooth; light gray	Smooth surface; light gray	Very smooth surface; light gray	Very smooth; bright gray	Very smooth: bright gray

250

Table 12-4 Metal Identification by Chips

	Copper	Brass and Bronze	Aluminium and Alloys	Monel Metal	Nickle	Lead
Appearance of chip	Smooth chips; saw edges where cut	Smooth chips; saw edges where cut	Smooth chips; saw edges where cut	Smooth edges	Smooth edges	Any shaped chip can be secured because of softness
Size of chip	Can be continuous if desired	Can be continuous if desired	Can be continuous if desired	Can be continuous if desired	Can be continuous if desired	Can be continuous if desired
Facility of chipping	Very easily cut	Easily cut; more brittle than copper	Very easily cut	Chips easily	Chips easily	Chips so easily it can be cut with penknife

Table 12-4 *(continued)*

	White Cast Iron	Gray Cast Iron	Malleable Iron	Wrought Iron	Low-Carbon Steel and Cast Steel	High-Carbon Steel
Appearance of chip	small broken fragments	Small, partially broken chips but possible to chip a fairly smooth groove	Chips do not break short 2.5 in cast iron	Smooth edges where cut	Smooth edges where cut	Fine grain fracture; edges lighter in color than low-carbon steel
Size of chip		1/8 in.	1/4-3/8 in.	Can be continuous if desired	Can be continuous if desired	Can be continuous if desired
Facility of chipping	Brittleness prevents chipping a path with smooth sides	Not easy to chip because chips break off from base metal	Very tough, therefore harder to chip than cast iron	Soft and easily cut or chipped	Easily cut or chipped	Metal is usually very hard, but can be chipped

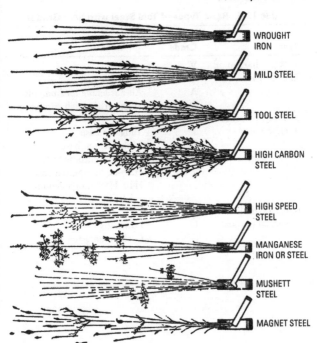

WROUGHT IRON

MILD STEEL

TOOL STEEL

HIGH CARBON STEEL

HIGH SPEED STEEL

MANGANESE IRON OR STEEL

MUSHETT STEEL

MAGNET STEEL

Fig. 12-1 Using the spark test for identifying metals.

TOOL STEELS

Tool steels are designed to provide wear resistance and toughness combined with high strength. They are basically high-carbon alloys. Chemistry provides the balance of toughness and wear desired for this type of steel.

Several classifications or breakdown systems have been applied to tool steels. See Table 12-5.

Table 12-5 Basic Types of Tool Steel with SAE Grades

Type	AISI-SAE Grade	
1. Water hardening	W	
2. Cold work	O	Oil-hardening
	A	Air-hardening medium-alloy
	D	High carbon/high chromium
3. Shock resisting	S	
4. High speed	T	Tungsten base
	M	Molybdenum base
5. Hot work	H	H1–H19: chromium base
		H20–H39: tungsten base
		H40–H59: molybdenum base
6. Plastic mold	P	
7. Special purpose	L	Low alloy
	F	Carbon–tungsten

13. REACTIVE AND REFRACTORY METALS

Special alloys of the reactive and refractory metals have been used extensively in the nuclear and aerospace industries, as well as for many other industrial purposes. The reactive metals include beryllium, titanium, and zirconium. The refractory metals include columbium, molybdenum, tantalum, tungsten, vanadium, and zirconium.

WELDING REACTIVE AND REFRACTORY METALS

Gas tungsten arc welding (GTAW), commonly known as TIG welding, is the most widely used process for welding refractory alloy sheets up to 1/8 inch thick. Either argon or helium (never carbon dioxide) is used as the inert shielding gas. Plasma arc welding (PAW), braze welding, and brazing can also be used for thicknesses up to 3/4 inch.

Note the following:

- Perfect fitup as well as a thorough cleaning of the metal surface is required immediately prior to welding.

- Clean the surface by degreasing, followed by a brushing with a steel brush and, if necessary, sandblasting. Sandblasting is used to remove scale too difficult for a metal brush to remove. Stainless steel brushes are recommended because other types may leave rust or metal deposits on the surface.

- Never expose the heated metal to the atmosphere. Protect it from atmospheric oxygen and other contaminants by shielding it in an inert gas chamber or a vacuum chamber.

- Construct a dry box around the area to be welded and pump inert shielding gas into the space. The inert gas chamber permits closer control of the atmospheric conditions around the weld.

- To ensure proper shielding when working outside a dry box, use a trailing shield of inert gas that meets the contours of the weldment. Note: An indication of inadequate shielding is a change in color of the electrode tip from bright to discolored.

- A flux is not required when welding reactive and refractory metals.

- A filler metal similar in composition to that of the base metal is used with certain thicknesses. Note: Proper joint fitup makes this unnecessary for thinner gauges.

- Weld slowly to avoid cracking. Note: Preheating can reduce weld cracking, but it must be done in a protective inert gas atmosphere.

Beryllium and Beryllium Alloys

Beryllium is a lightweight reactive metal characterized by extreme hardness (it can scratch a glass surface) and a color similar to that of magnesium. It has a melting point of 2332°F (1285°C), a density comparable to that of magnesium, and a high electrical and thermal conductivity. It is frequently used as an alloying element with other metals (e.g., copper, nickel, and magnesium) for increased strength, elasticity, and other characteristics. Its light weight (it is lighter than aluminum), tensile strength (about 55,000 psi), and relatively high melting temperature have suited it to applications in the aerospace industries. Other applications include the use of beryllium wire in the production of electrical circuits.

NOTE

Beryllium has a lower melting point than the refractory metals (tungsten, molybdenum, tantalum, and columbium). On the other hand, it exhibits many of the same characteristics as refractory metals, especially its need to be shielded from the contaminating effects of oxygen and other gases at high temperatures.

Beryllium is derived primarily from beryllium aluminum silicates. The beryllium metal is prepared by electrolysis. Beryllium (1.0 to 2.5 percent) is combined with nickel (up to 1.0 percent) and copper to form the beryllium-copper alloy. This alloy has a relatively high tensile strength, up to 180,000 psi.

Beryllium is also used in small amounts (0.2 to 0.25 percent) as an alloy in the production of a cast beryllium bronze. Iron, silicon, and cobalt are sometimes added to the beryllium-copper alloy to achieve various desired characteristics.

CAUTION

Beryllium dust and fumes are very toxic. Their extreme danger rests on the fact that there is no known cure (only remission agents) for their effects. In extreme cases, breathing beryllium dust and fumes can be fatal. *Every safety precaution should be taken when working with beryllium and beryllium alloys.*

Beryllium can be joined by braze welding and brazing. High-temperature welding processes are not recommended.

Note the following:

- A low-temperature filler metal such as aluminum silicon or a silver-base alloy is recommended.

- The filler metal should be placed in the joint before braze welding or brazing.

- Keep braze welding or brazing time as short as possible to prevent overheating the surface.

Columbium and Columbium Alloys

Columbium (or *niobium)* resembles steel in appearance, and is closely associated with tantalum in properties. It has a yellowish-white color and a melting point of 4474°F (2468°C). Its density is slightly greater than that of iron. The tensile strength of columbium is 48,000 to 59,000 psi in the annealed condition, and up to 130,000 psi as drawn wire.

Columbium exhibits excellent strength, ductility, and corrosion-resistance. However, its resistance to corrosion is somewhat limited by its strong reaction to oxides. Oxidation becomes a serious problem at temperatures in excess of 400°C, necessitating the use of an inert shield when working the metal at higher temperature.

Columbium is a useful alloying element in other metals. It is particularly useful in imparting stability to stainless steel. Because of its low resistance to thermal neutrons, columbium is frequently used in nuclear equipment.

Columbium is added to steels in the form of ferro-columbium (50.0 to 60.0 percent columbium, 33.0 to 43.0 percent iron, 7.0 percent silicon). Columbium (in amounts exceeding 75.0 percent) is alloyed with such metals as tungsten, zirconium, titanium, and molybdenum to produce columbium alloys that will retain high tensile strength at high temperature (in the 2000 to 5000°F range).

Columbium is a refractory metal, a characteristic that encourages frequent use of columbium alloys in the production of high-temperature-resisting components for missiles, turbines, and jet engines.

Gas tungsten arc welding (GTAW) performed in a vacuum chamber is the method commonly used to join columbium and the columbium alloys. Braze welding and brazing can also be used to join columbium.

Note the following:

- Preheating is required for some columbium alloys in order to avoid weld cracks.

- Heat treatment should be performed in an inert shielding gas or a vacuum chamber.

- A fast welding speed with minimum amp input produces the best penetration.

- Copper backing bars should be used to extract heat from the weld.

Molybdenum and Molybdenum Alloys

Molybdenum resembles steel in color, but most of its physical properties are like those of tungsten. It has a melting point of 4730°F (2610°C), which qualifies it as one of the refractory metals (those having a melting point above 3600°F) along with columbium (niobium), tungsten, and similar metals. Molybdenum is a very important alloying element in the production of iron and steel.

As an alloying element, molybdenum is not generally used in quantities exceeding 4 percent. Research in the development of molybdenum steels began in the 1890s, but it was not until after World War I that it was found possible to produce such steel on an economical basis. It received widespread use in the automotive industry, and its application to other areas expanded rapidly. Molybdenum is now widely used in the production of tool steels and high-speed steels, where it contributes to wear resistance, hardness, and strength. Molybdenum is also added to gray iron (in amounts less than 1.25 percent) to increase its tensile strength and harden ability.

Molybdenum is added to steel to increase hardness, endurance, corrosion resistance (as in stainless steel), and the tendency toward deep hardening. Because of the high melting point of molybdenum and its alloys, they are employed in the manufacture of rocket and gas turbine engines. Its greatest use, however, is in the electronics and nuclear industries.

The gas tungsten arc welding (GTAW) and gas metal arc welding (GMAW) processes, both employing inert shielding gases, are used to weld molybdenum and its alloys. Brazing also can be used to join the thinner gauges of molybdenum.

Note the following:

- Use argon or helium as the inert shielding gas. Note: *Never* use carbon dioxide as a shielding gas when welding molybdenum.

- Preheating is usually not necessary when welding molybdenum.

- When brazing, be sure to select a brazing filler metal that is suitable for the application. Temperatures for brazing filler metals range from 1200 to 4500°F.

- Keep the brazing time as short as possible to minimize heat buildup.

Tantalum and Tantalum Alloys

Tantalum is a refractory metal in the same category as tungsten, molybdenum, and columbium. It has a color ranging from whitish to silvery gray, a tensile strength of 50,000 psi, and a melting point of 5425°F. It is characterized by an unusually strong resistance to acids and corrosion. For this reason, it finds useful application in the manufacture of chemical and surgical equipment.

Tantalum and its alloys have many other interesting applications, including their use as filaments in electric light bulbs and components in jet engines and rocket motors. Because tantalum is a very ductile metal, it can be drawn or rolled without annealing.

Tantalum has an extremely strong reaction to oxygen in the surrounding atmosphere or in the oxygen cutting stream of an oxyacetylene cutting torch. For this reason, tantalum loses much of its effectiveness as a high-temperature structural metal. Commercially pure tantalum is soft and ductile.

A thin oxide layer, similar to that found on aluminum, covers the surface of tantalum and contributes to the metal's resistance to acids.

Gas tungsten arc welding (GTAW) is the most commonly used welding process for joining tantalum and its alloys. Brazing is also used in certain applications.

Note the following:

- Shield the weld metal and surrounding area with an appropriate inert gas.

- Copper brazing in an inert gas vacuum can be used to join thin gauges of tantalum.
- The welding characteristics of tantalum are similar in many respects to those described for columbium.

Titanium and Titanium Alloys

Titanium is characterized by a silvery color, a high resistance to corrosion, and the highest affinity for carbon of all known metals. As a result of this carbon affinity, titanium functions as an alloying element (less than 1 percent) in steels to stabilize the carbon and prevent cracking.

Another characteristic of titanium is its strong tendency to form carbides. For this reason, it is used in some chromium steels to counteract chromium depletion tendencies.

Titanium has a melting point of 3035°F (1800°C). The titanium alloys are as strong as steel, but are 50 to 60 percent lighter in weight. Commercially pure titanium has a tensile strength of 45,000 psi, but the hardened titanium alloys may have tensile strengths extending to 200,000 psi. The ductility of this metal is relatively low.

Perfect fitup and a thorough cleaning of the metal surface are required prior to welding. The surface may be cleaned by degreasing, followed by steel brushing and sandblasting. Stainless steel brushes are recommended, because other types may leave rust or metal deposits on the surface. Sandblasting is used to remove scale that is too difficult for the metal brush. All cleaning must be done immediately prior to welding, and the surface must be *thoroughly* cleaned of all contaminants.

Because of its strong as an absorbent for most solids and gases (except argon and helium), titanium must be welded under specially shielded conditions. In addition to the use of argon or helium as the shielding gas in gas tungsten arc welding, a trailing shield should also be used until the surface has cooled to approximately 750°F.

The gas tungsten arc welding (GTAW), gas metal arc welding (GMAW), plasma arc welding (PAW), braze welding, and brazing processes can all be used to join all thicknesses of titanium and the titanium alloys. Oxyacetylene welding is not recommended for welding this metal and its alloys.

Note the following:

- In addition to the use of argon or helium as the shielding gas when using the gas tungsten arc welding (GTAW) process, a trailing shield should also be used until the surface has cooled to approximately 750°F.

- Filler metals similar in composition to that of the base metal are used with certain thicknesses of titanium. Proper joint fitup makes this unnecessary for thinner gauges of titanium and titanium alloys.

- A flux is not required.

Tungsten and Tungsten Alloys

The color of tungsten ranges from steel gray to a silvery white. It has a melting point of 6039°F (3337°C), the highest of all metals. As a result of this characteristic, tungsten has found applications in the aerospace industry for rocket components that must withstand extreme temperatures. Tungsten is also used in the electronics industry, in welding, in the electrical industry, and in the production of high-speed cutting tools.

Tungsten is an important alloy in the production of steels. Important among these steels are the tungsten tool steels (up to 20.0 percent tungsten), and particularly the high-speed tool steels. As an alloying element of steel, tungsten contributes characteristics such as hardness, strength, and wear resistance.

Note the following:

- Tungsten and tungsten alloys have weldability characteristics similar to those described for molybdenum and its alloys.

- Torch brazing also can be used to join tungsten and its alloys.

- Do not use a nickel-base filler metal when brazing tungsten. The interaction between nickel and tungsten can cause weld problems.

Vanadium and Vanadium Alloys

Vanadium is a grayish-white metal used primarily in steel production and as an alloying element for other metals. Vanadium tends to increase the hardenability of a metal and to reduce or eliminate the harmful effects of overheating.

Vanadium melts at a temperature of 3236°F. Because it is not subject to oxidation (it is highly resistant to corrosion), it serves as a strong deoxidizer in steels. It increases the tensile strength, has little or no effect on the ductility of the metal, and it greatly increases grain growth. In the annealed condition, vanadium has a tensile strength of 66,000 psi.

Some steels in which vanadium is present as a major alloying element are Python steel (0.25 percent vanadium), Vasco vanadium (0.20 percent), and SAE 6145 steel (0.18 percent). These and other (Colonial No. 7, Elvandi, etc.) are referred to in general as *vanadium steels*. This general grouping is frequently divided into subgroups (e.g., *chromium-vanadium steels* or *carbon-vanadium steels*) depending on the alloys present.

Note the following:

Use the same welding tips listed for zirconium and the zirconium alloys to weld vanadium and its alloys. The weldability of both metals is very similar.

Zirconium and Zirconium Alloys

Zirconium is a silvery-white metal with a melting point of 3330°F (1700°C), or almost three times that of aluminum. It has metallurgical characteristics very similar to those of titanium. Pure zirconium has a relatively low tensile strength of approximately 32,000 psi. It exhibits a strong corrosion resistance and excellent thermal stability.

A principal application of zirconium is as an absorber of gases in electronic tubes. It is also used as an alloying element in steel to prevent brittleness and age hardening. Other applications include zirconium sheets, foil, wire, and rod; surgical tools and components; and chemical equipment. Its thermal stability and resistance to corrosion make it highly suited for construction in the nuclear field.

Zircaloy-2 and *Zircaloy-3* are two of the better known trade names for zirconium alloys. *Zircaloy-2*, a product of Westinghouse Electric Corporation, contains 98.28 percent zirconium. The other elements (tin, iron, chromium, nickel) are added to increase its strength and to reduce the tendency of zirconium to absorb hydrogen. Both *Zircaloy-2* and *Zircaloy-3* have excellent weldability.

Gas tungsten arc welding (GTAW) or gas metal arc welding (GMAW) are the methods commonly used to join zirconium and the zirconium alloys.

Note the following:

- Zirconium reacts to both oxygen and nitrogen, which makes it necessary to weld in an inert gas shielding atmosphere.

- Critical welds should be performed inside a dry box.

- Maintain shielding until the weld metal and the heated affected zone has cooled to at least 700°F.

- The rate of cooling can be safely increased by using copper backup plates and clamps.

- GMAW welding wire must be of the hard-drawn type.

14. GALVANIZED METALS

Galvanized iron and steel are ferrous metals coated with a thin layer of zinc for protection against corrosion. Galvanized steel is widely used in the building construction industry for siding and roofing.

Any of the fusion welding processes can be used to join galvanized iron or steel. Shielded metal arc welding (SMAW) is probably the most widely used process for welding galvanized metals (see Table 14-1).

CAUTION

The heat of the arc or flame liberates zinc fumes during welding. Provision must be made to protect the welder from these fumes by providing adequate ventilation if the welding is being done inside, and by wearing the necessary respirator protection. This precaution should be taken at all times and for every job involving galvanized metals. If the welder should experience any nausea during or after welding galvanized material, medical help should be sought immediately.

Note the following:

- The American Welding Society advises removing the zinc coating on galvanized steels before welding, brazing, or soldering. Any heat applied to the zinc coating will release potentially harmful zinc fumes into the atmosphere. Remove 2 to 4 inches of the zinc coating on both sides of the intended weld line and on both sides of the workpiece. Grinding away the zinc coating is the preferred method.

- Burning away the zinc coating from the weld zone is an alternative to grinding the surface.

- Note: Removing the zinc coating from the weld zone is not an absolute requirement. Check the job specifications for guidance.

Table 14-1 Some Recommendations for Welding and Joining Galvanized Metals

Welding or Joining Process	Comments
Shielded Metal Arc Welding (SMAW)	• Use flux-covered electrodes when welding galvanized metals with the SMAW process.
	• Reduce the electrode angle by about 30° and whip it back and forth to push the molten zinc pool away from the weld. Note: The electrode angle and whipping motion result in a reduced welding speed.
	• Use a wider than normal root opening to facilitate penetration.
	• Remove the spatter after the weld is finished. Slightly more spatter is produced when using the SMAW arc to remove the zinc coating than is the case when welding an uncoated surface.
Gas Metal Arc Welding (GMAW)	• GMAW is commonly recommended for welding thinner gauges of galvanized metals.
	• The weld's mechanical properties are unaffected by the zinc coating.
	• The GMAW process requires a higher heat input and lower welding speeds for zinc coated materials. Both are required to burn off and remove the coating.
	• There may be a reduction in welding speed if the zinc has not been removed, because the galvanized coating must be burned off ahead of the weld.

(continued)

Table 14-1   (continued)

Welding or Joining Process	Comments
	• Use a 100-percent CO_2 gas to shield the weld.
	• The arc is stable, but penetration is less than for uncoated metals.
Oxyacetylene Welding (OAW)	• Use the same procedure recommended for welding uncoated steel.
	• Move the filler rod back and forth to produce a ripple weld.
	• Because low travel speed is necessary to bring the joint edges to the fusion temperature, the extra heat causes the zinc coating's appearance to be affected over a much greater area than with other welding processes.
Braze Welding	• Braze welding (as well as brazing) is used where full-strength welds are not required, where a color match between the weld material and the base metal is not necessary, or where high temperatures cannot be used.
	• No preheating is required when braze welding steel.
	• Use a deoxidizing flux when braze welding galvanized metals. A deoxidizing flux prevents atmospheric oxygen from attacking the zinc coating while it is in the molten state.

(continued)

Table 14-1 *(continued)*

Welding or Joining Process	Comments
	• When braze welding galvanized pipe, be sure to coat both the inside and outside of the pipe with the liquid flux about 3 inches back from the joint. When the deposited metal cools, the contraction caused by the cooling draws the ends of the pipe together.
Brazing	• Coat both sides of the galvanized sheets along the seam with a thick coating of a paste flux. This paste coating protects the galvanized coating on the metal.
	• Braze with a low-temperature brazing rod or a nickel brazing rod together with a matching flux. Do *not* use a bronze rod with a higher-temperature melting point. A higher-temperature rod will burn the galvanized coating.
Soldering	• Used for joining furnace ductwork, gutters, and downspouts.
	• Electro-galvanized steel sheet is easier to solder than steel sheet covered with hot-dipped zinc coatings.
	• Chromate treatment of zinc-coated sheets may interfere with solder flow.

15. SOFT METALS WELDING

ALUMINUM

Aluminum is a lightweight metal with one-third the weight of steel. Aluminum alloy 7178-T6 is heat treatable and can develop strengths up to 88,000 psi. With this strength-to-weight ratio, aluminum has an advantage over steels, especially in aerospace applications.

Aluminum has excellent forming characteristics. It can be bent, extruded, joined, or machined with comparative ease. Aluminum is an excellent conductor of electricity and makes a good heat conductor material, with the added advantage of having high atmospheric corrosion resistance.

Metallurgists have developed many alloys of aluminum for special products and other applications. Aluminum alloys are:

- Aluminum-copper
- Aluminum-manganese
- Aluminum-silicon
- Aluminum-magnesium
- Aluminum-silicon-magnesium
- Aluminum-zinc

CLASSIFICATION OF WROUGHT-ALUMINUM ALLOYS

A classification system for wrought-aluminum alloys was established by the Aluminum Association in October, 1954. The standardized system consisted of a four-digit number. The first number indicates the alloy group, the second indicates the impurity limit in the original alloy, and the last two identify either the specific alloy or the purity of the aluminum. See Table 15-1.

This numbering system for example, would be for an aluminum alloy classified as 7178-T6. The first digit (7) indicates

Table 15-1 Classification of Wrought Aluminum Alloys

Aluminum Association Classification Number	Alloy Group
1xxx	Aluminum 99.00% Min.
2xxx	Copper
3xxx	Manganese
4xxx	Silicon
5xxx	Magnesium
6xxx	Magnesium and Silicon
7xxx	Zinc
8xxx	Other Elements
9xxx	Unused Series

that the alloy is of the zinc group. The second digit (1) indicates the original alloy modifications and may use numbers 1 through 9 to show alloy modifications or impurity limits of the alloy group. The last two numbers (7 and 8) express the minimum aluminum percentage to the nearest 0.01 percent. The major alloying element is aluminum, thus the 7178 alloy has the purity of 99.78 percent aluminum purity, with control over alloy modification at level 1. Zinc is the alloy group being used.

A dash separates the numbers from the letter T and indicates the condition of the alloy.

The alloy illustrated before was 7178-T6, which means that the metal which can withstand 88,000 pounds per square inch (psi) was solution-treated and artificially aged. The metal had a specific heat treatment to provide this exceptional strength. See Table 15-2.

Zinc

Zinc can be used as a mixer to produce an aluminum alloy. As a metal with a low melting point, it can be cast easily at temperatures ranging from 750° to 800° F. Because

Table 15-2 Uses of Wrought-Aluminum Alloys

1060	Chemical equipment, railroad tank cars
3003	Ductwork, truck panels, architectural application, builders' hardware
3004	Hydraulic tubing for commercial vehicles, storage tanks, roofing
5052	Bus and truck bodies, aircraft tubing, kitchen cabinets, marine hardware
5454	Welded structures, saltwater applications
5456	Deck housing, heavy-duty structures, overhead cranes
5457	Auto and appliance trim
6061	Transportation equipment, heavy-duty structures, marine, pipe, furniture, bridge rails
7078	Structural aircraft parts

zinc can be cast under great pressure at these low temperatures, it produces a very accurate and consistent product.

The machinability of aluminum-base alloys varies greatly, although most cast alloys can be machined easily. For the wrought alloys, with the exception of a few special types, special tools and techniques are desirable if large-scale machining is to be done.

WELDING ALUMINUM

Aluminum is difficult to weld using the oxyacetylene process due to its low melting point. Aluminum melts at about $1220°F$ and gives no warning as it approaches the melting point temperature. MIG welding was developed specifically for working with aluminum.

Aluminum can be welded with oxyacetylene, but the process requires a skilled welder.

1. Begin by preheating the metal with an excess acetylene flame. This flame deposits carbon on the surface, which looks black.

2. Next, heat the joint area with a neutral flame. Do this until the carbon is removed.

3. At this point, apply some special liquid aluminum welding flux.

4. While keeping the cone of the flame at least 1 inch above the joint, introduce a special aluminum welding rod into the flame (the object being to tin the joint with the aluminum rod).

5. After the joint has been tinned, melt more rod onto the area and build up the joint. The flame of the torch *must not* be played on the joint, as this may cause the aluminum to become very weak.

6. After the weld has been completed, all flux must be removed from the joint and surrounding area. This is commonly done with warm water and a brush.

Aluminum Filler Materials

There are a number of variables that need to be considered during the selection of the most suitable filler alloy for a particular base alloy and component operating condition. When choosing the optimum filler alloy, both the base alloy and the desired performance of the weldment are prime considerations. What is the weld subjected to, and what is it expected to do?

Aluminum alloys, with their various tempers, constitute a wide and versatile range of manufacturing materials. To achieve the best product design and the most successful welding, it is important to understand the differences among the many alloys available and their various performance and weldability characteristics. In developing arc welding procedures for different alloys, always consider the specific alloy being welded. See Table 15-3.

Table 15-3 Designations of Aluminum Alloy Conditions

— F:	As fabricated
— O:	Annealed, recrystallized
— H:	Strain hardened
— W:	Solution treated, unstable temper
— T:	Heat treated to stable temper
— T3:	Solution treated and cold-worked
— T4:	Solution treated
— T5:	Artificially aged only
— T6:	Solution treated and artificially aged
— T7:	Solution treated and stabilized
— T8:	Solution treated, cold-worked, and artificially aged
— T9:	Solution treated, artificially aged, and cold-worked
— T10:	Artificially aged and cold-worked

If an alloy has been strain-hardened or cold-worked, these
   designations may be used.

— H1:	Strain-hardened only
— H2:	Strain-hardened and partially annealed
— H3:	Strain-hardened and stabilized

ALUMINUM SOLDERING
Chapter 7 covers the soldering of many metals, including aluminum. Fluxes and heats, as well as the various soldering procedures, are covered in detail.

COPPER AND COPPER ALLOYS
Copper and copper alloys are among the most important engineering materials because of several good properties:

- Electrical and thermal conductivity
- Corrosion resistance
- Metal-to-metal wear resistance
- Distinct, aesthetic appearance

Copper and most copper alloys can be joined by welding, brazing, and soldering.

SURFACE PREPARATION

The weld area should be clean and free of oil, grease, dirt, paint, and oxides prior to welding. Wire brushing with a bronze wire brush should be followed by degreasing with a suitable cleaning agent. The oxide film formed during welding should also be removed with a wire brush after each weld run is deposited.

Pre-heating

Welding thick copper sections requires a high preheat, because copper conducts heat rapidly from the weld joint into the surrounding base metal. Most copper alloys—even in thick sections—do not require preheating, because their thermal diffusivity is much lower than for copper. To select the correct preheat for a given application, consideration must be given to the welding process, the alloy being welded, the base metal's thickness, and to some extent the overall mass of the weldment. Aluminum-bronze and copper-nickel alloys should not be preheated. It is desirable to limit the heat to as localized an area as possible to avoid bringing too much of the material into a temperature range that will cause a loss in ductility. It is also important to ensure that the preheat temperature is maintained until welding of the joint is completed.

GAS METAL ARC WELDING (GMAW) OF COPPER AND COPPER ALLOYS

ERCu copper electrodes are recommended for GMAW of copper. Deoxidized copper is a versatile 98-percent pure copper alloy for the GMAW of copper. The gas mixture required will be largely determined by the thickness of the copper section to be welded. Argon is generally used for thicknesses of 6 mm or less.

The helium-argon mixtures are used for welding thicker sections. The filler metal should be deposited with stringer beads or narrow weave beads, using spray transfer. Table 15-4 gives general guidance on procedures for GMAW of copper.

Recommended Shielding Gases for the GMA Welding of Copper and Copper Alloys—Welding-Grade Argon

- Ar + >0–3% O_2 or equivalent shielding gases
- Ar + 25% He or equivalent shielding gases
- He + 25% Ar or equivalent shielding gases

GMAW OF COPPER-SILICON ALLOYS

ERCuSi-A type welding consumables plus argon shielding and relatively high travel speeds are used with this process. Silicon bronze is a copper-based wire recommended for GMAW of copper-silicon alloys. It is important to ensure that the oxide layer is removed by wire brushing between passes. Preheat is unnecessary, and interpass temperature should not exceed 100°C.

GMAW OF COPPER-TIN ALLOYS (PHOSPHOR BRONZE)

These alloys have a wide solidification range that creates a coarse, dendritic grain structure. Care must be taken during welding to prevent cracking of the weld metal. Hot-peening of the weld metal will reduce the stresses developed during welding, and the likelihood of cracking. The weld pool should be kept small by using stringer beads at high travel speed.

GAS TUNGSTEN ARC WELDING (GTAW) OF COPPER AND COPPER ALLOYS

Gas Tungsten Arc Welding of Copper

Copper sections up to 16 mm in thickness can be successfully welded using the gas tungsten arc welding process. The recommended filler wire is a filler metal whose composition is similar to that used for gas tungsten arc welding (GTAW) of copper and copper alloy sections up to 1.6 mm thick.

Table 15-4 Suggestions for GMAW of Copper

Metal Thickness	Electrode Diameter	Preheat# Temperature	Welding Current	Voltage Rate	Gas Flow Rate(L/min)	Travel Speed
1.6 mm	0.9 mm	75°C	150–200	21–26	10–15	500 mm/min
3.0 mm	1.2 mm	75°C	150–220	22–28	10–15	450 mm/min
6.0 mm	1.2 mm	75°C	180–250	22–28	10–15	400 mm/min
6.0 mm	1.6 mm	100°C	160–280	28–30	10–15	350 mm/min
10 mm	1.6 mm	250°C	250–320	28–30	15–20	300 mm/min
12 mm	1.6 mm	250°C	290–350	29–32	15–20	300 mm/min
16 mm+	1.6 mm	250°C	320–380	29–32	15–25	250 mm/min

Over, 1.6 mm thick sections, argon shielding gas is preferred. Helium mixes are preferred for welding sections over 1.6 mm thick. In comparison to argon, argon-/helium mixes permit deeper penetration and higher travel speeds at the same welding current. A mixture containing 75 percent% helium and −25 percent% argon mixture is commonly used because of the for good penetration characteristics of helium and combined with the easy arc starting and improved arc stability properties of.

Forehand welding is preferred for the gas tungsten arc welding of copper with stringer beads or narrow weave beads. Typical conditions for manual GTAW of copper are shown in Table 15-5.

BRAZING COPPER AND COPPER ALLOYS

The principle of brazing is to join two metals by fusing with a filler metal. The filler metal must have a lower melting point than the base metals but greater than 450°C (use of a filler metal with a melting point less than 450°C is called soldering). The filler metal is usually required to flow into a narrow gap between the parts by capillary action.

Brazing is used widely to join copper and copper alloys, except for aluminum bronzes with more than 10 percent aluminum and alloys with more than 3 percent lead.

Where brazing of copper is useful:

- Electrical manufacturing industry
- Building mechanical services
- Heating installations
- Ventilation devices
- Air-conditioning equipment

To achieve an adequate bond during brazing, the following points should be considered:

- Make joint surfaces clean and free of oxides, and so forth.

Table 15-5 Manual GTAW Conditions for Copper (Aufhouser)

Metal Thickness (mm)	Shielding Gas	Tungsten Type & Welding Current	Welding Rod Diameter	Preheat Temperature	Welding Current
0.3–0.8	Argon	Thoriated/DC–	—	—	15–60
1.0–2.0	Argon	Thoriated/DC–	1.6 mm	—	40–170
2.0–5.0	Argon	Thoriated/DC–	2.4–3.2 mm	50°C	100–300
6.0	Argon	Thoriated/DC–	3.2 mm	100°C	250–375
10.0	Argon	Thoriated/DC–	3.2 mm	250°C	300–375
12.0	Argon	Thoriated/DC–	3.2 mm	250°C	350–420
16.0	Argon	Thoriated/DC–	3.2 mm	250°C	400–475

- Provide for the correct joint gap for the particular brazing filler metal.
- Establish the correct heating pattern so that the filler metal flows up the thermal gradient into the joint.

Surface Preparation

Standard solvent or alkaline degreasing procedures are useful in the cleaning of copper-base metals. Be careful when using mechanical methods to remove surface oxides. Chemical removal surface oxides can be removed chemically by using an appropriate pickling solution such as Chrome Bright.

Joint Design Considerations

The distance between the members to be joined must be controlled within certain tolerances that depend on the brazing alloy and the parent metal used. The best joint gap typically is from 0.04 to 0.20 mm.

Brazing of Copper and Copper Alloys

Usually a joint overlap of three or four times the thickness of the thinnest member to be joined will work. The goal is to use as little material as possible to achieve the desired strength. Figure 15-1 illustrates common joint design for silver brazing.

Flame Adjustment

Use a neutral flame (produced by mixing equal amounts of oxygen and acetylene at the same rate). The white inner cone of the flame is clearly defined and shows no haze.

Flux Removal

Flux residue must be removed by using one of the following methods:

- A dilute solution of hot caustic soda dip
- Wire brushing and rinsing with hot water
- Wire brushing and steam

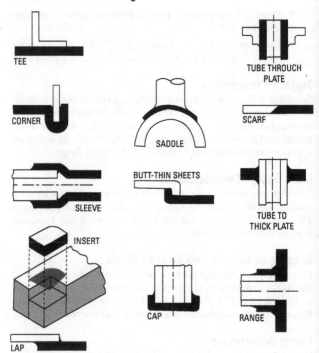

Fig. 15-1 Common joint design for silver brazing. (Aufhauser)

Incomplete flux removal may cause weakness and failure of the joint.

BRAZE WELDING OF COPPER

Braze welding and fusion welding are similar, except braze welding uses a filler metal of lower melting point than the parent metal. The braze-welding process gets its strength from the tensile strength of the filler metal deposited and the bond strength developed between the filler metal and the parent metal. Oxyacetylene is usually preferred because of its easier flame setting and rapid heat input.

Choice of Alloy
The job requirement determines the choice of alloy. It is also determined by the strength required in the joint, resistance to corrosion, operating temperature, and economic considerations. Low-fuming bronze and flux-coated low-fuming bronze are commonly used alloys.

Joint Preparation
For typical joint designs, see Figure 15-2.

Flame Adjustment
Use a slightly oxidizing flame.

Flux
Use copper and brass flux. Mix to a paste with water and apply to both sides of the joint. The rod can be coated with paste or heated and dipped in dry flux.

Preheating
Preheating is recommended for heavy sections only.

Blowpipe and Rod Angles
Keep the blowpipe tip at a 40 to 50° angle to the metal surface. The distance of the inner cone from the metal surface should be 3.25 mm to 5.00 mm. The angle between the filler rod and the metal surface should be 40 to 50°. See Table 15-6.

After preheating, or after the joint is raised to a temperature sufficient to permit alloying of the filler rod and copper, melt a globule of metal from the end of the rod and deposit it into the joint to *wet*, or *tin*, the surface. When tinning occurs, begin welding with the forehand technique (see Figure 15-3). Do not drop filler metal on untinned surfaces.

DOUBLE VEE JOINT
(THICKER THAN 13mm)

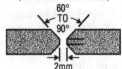

60°
TO
90°

2mm

LAP JOINT

13mm
min.

BELL TYPE BUTT JOINT

t

45°

d

D

(D MUST NOT BE
LESS THAN d + dt)

BUTT JOINT

t

6t

½t

SINGLE VEE JOINT
(3mm TO 6mm THICK WITH
SHARP CONNERS REMOVED)

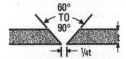

60°
TO
90°

¼t

DIMINISHING JOINT

3mm TO 5mm

SINGLE VEE JOINT
WITH ROOT FACE
(THICKER THAN 6mm)

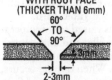

60°
TO
90°

2-3mm

STUB BRANCH JOINT

30° mm

Fig. 15-2 Joint designs for braze welding. (Aufhouser)

**Table 15-6 Filler Rod and Tip Size for Braze
Welding of Copper**

Plate Thickness (mm)	Filler Rod (mm)	Blowpipe Acetylene consumption (Cu. L/Min)	Tip Size
0.8	1.6	2.0	12
1.6	1.6	3.75	15
2.4	1.6	4.25	15
3.2	2.4	7.0	20
4.0	2.4	8.5	20
5.0	3.2	10.0	26
6.0	5.0	13.5	26

Flux Removal

Any of the following methods may be used to remove flux
residue:

- A grinding wheel or wire brush and water
- Sand blasting
- A dilute caustic soda dip

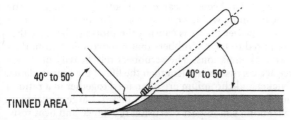

Fig. 15-3 Braze welding forehand technique. (Aufhouser)

NICKEL

Nickel is a versatile metal. In pure form it is used in the electroplating industry to coat metals that will be exposed to corrosive atmospheres, sea water, sulfuric acid, brine, and caustic solutions.

Nickel is added to other metals for its ablity to increase toughness and impact resistance, especially at low temperatures. It also:

- Lessens distortion in quenching
- Improves corrosion resistance
- Lowers the critical temperature
- Widens the temperature range for successful heat treatment

Nickel is added in amounts of 2 to 5 percent, often combined with other alloying elements to improve toughness. Using 12 to 20 percent nickel in steel with low carbon content provides good corrosion resistance. A steel with 36 percent nickel has a thermal expansion coefficient of almost zero. This alloy is commonly known as Invarn and is used for measuring devices.

Because of its high cost, nickel should only be used where it is uniquely effective, as in providing low-temperature impact resistance for cryogenic steels. A metric ton of nickel can sell for $30,000 in 2006, and probably more in following years. It has become scarce, and both deeper digging and chemical mining will have to be done. The demand for more coal-fired power plants is causing the shortage, because they are required to have scrubbers that remove sulfur from their smokestack emissions. The scrubbers rely heavily on nickel to reduce corrosion and lengthen the life of the power plants.

Nickel and chromium are often used together in a ratio of about 1 part chromium to 2 parts nickel. Large percentages of chromium can impart corrosion resistance and heat resistance, but in the amounts used in low-alloy steels the effects are minor. Less than 2 percent chromium usually is required. Chromium carbides are often desirable for their superior wear resistance.

NICKEL-BASED ALLOYS

Nickel-based alloys are noted for their strength and corrosion resistance, especially at high temperatures. Monel metal, found in an alloy form in Canada, contains about 67 percent nickel and 30 percent copper. It has been used for years in the chemical and food-processing industries because of its outstanding corrosion resistance. It probably has better overall corrosion resistance to more media than any other alloy. Monel metal is particularly resistant to saltwater corrosion, sulfuric acid, and even high-velocity, high-temperature steam. It has been used for steam turbine bladesbecause of its stem resistance. It can be polished for stainless steel sinks and artwork . It shines easily and has an excellent appearance; in fact, it is often used for ornamental trim and household ware. In its common form, Monel has a tensile strength of from 480 to 1170 MPa (70 to 170 ksi)ksi , depending on the amount of cold working. The elongation in 2 inches varies from 50 to 2 percent.

There are three special grades of Monel that contain small amounts of added alloying elements:

- **K Monel** contains about 3 percent aluminum and can be precipitation-hardened to a tensile strength of 1100 to 1240 MPa (160 to 180 ksi).

- **H Monel** has 3 percent silicon added.

- **S Monel** contains 4 percent silicon.

These special Monel grades are used for castings, and they can be precipitation-hardened. To improve the machining charactristics of Monel, a special free machining alloy known as **R Monel** is produced with about 0.35 percent sulfur.

Another use for nickel-based alloys is as electrical resistors. Mixed with chromium, nickel becomes *nichrome* and is used for heating elements in many electrical devices.

Table 15-7 Some Nonferrous Alloys for High-temperature Service

Alloy	C	Mn	Si	Cr	Ni	Co	Mo	W	Cb	Ti	Al	B	Zr	Fe	Other
Nickel base															
Hastelloy X	0.1	1.0	1.0	21.8	Balance	2.5	9.0	0.6	—	—	—	—	—	18.5	—
IN-100	0.18	—	—	10.0	Balance	15.0	3.0	—	—	4.7	5.5	0.014	0.06	—	1.0 V
Inconel 601	0.05	0.5	0.25	23.0	Balance	—	—	—	—	—	1.4	—	—	14.1	0.2 Cu
Inconel 718	0.04	0.2	0.2	19.0	Balance	—	3.0	—	5.0	0.9	0.5	—	—	18.5	0.2 Cu
M-252	0.15	0.5	0.5	19.0	Balance	10.0	10.0	—	—	2.6	1.0	0.005	—	—	—
Rene 41	0.09	—	—	19.0	Balance	11.0	10.0	—	—	3.1	1.5	0.01	—	—	—
Rene 80	0.17	—	—	14.0	Balance	9.5	4.0	4.0	—	5.0	3.0	0.015	0.03	—	—
Rene 95	0.15	—	—	14.0	Balance	8.0	3.5	3.5	3.5	2.5	3.5	0.01	0.05	—	—
Udimer 500	0.08	—	—	19.0	Balance	18.0	4	—	—	3	3	0.005	—	0.5	—
Udimer 700	0.07	—	—	15.0	Balance	18.5	5.0	—	—	3.5	4.4	0.025	—	0.5	—
Waspaloy B	0.07	0.75	0.75	19.5	Balance	13.5	4.3	—	—	3.0	1.4	0.006	0.07	2.0	0.1 Cu
Iron-nickel base															
Illium P	0.20	—	—	28.0	8.0	—	2.0	—	—	—	—	—	—	Balance	3.0 Cu
Incoloy 825	0.03	0.5	0.2	21.5	42.0	—	3.0	—	—	0.9	0.1	—	—	30	2.2 Cu
Incoloy 901	0.05	0.4	0.4	13.5	42.7	—	6.2	—	—	2.5	0.2	—	—	34	—
16-25-6	0.08	1.35	0.7	16.0	25.0	—	6.0	—	—	—	—	—	—	Balance	0.15 N
Cobalt base															
Haynes 150	0.08	0.65	0.75	28.0	—	Balance	—	—	—	—	—	—	—	20.0	—
MAR-M322	1.00	0.10	0.1	21.5	—	Balance	—	9.0	—	0.75	—	—	2.25	—	4.5 Ta
S-816	0.38	1.20	0.4	20.0	20.0	Balance	4.0	4.0	4.0	—	—	—	—	4.0	—
WI-52	0.45	0.5	0.5	21.0	1.0	Balance	—	11.0	2.0	—	—	—	—	2.0	—

Most of the nickel alloys are somewhat difficult to cast, but they can be forged and hot-worked. The heating usually must be done in controlled atmospheres. This is to avoid intercrystalline embrittlement.

Some of the nonferrous alloys are listed in Table 15-7. Note the many alloys containing nickel.

16. MAGNESIUM AND MAGNESIUM ALLOYS

Magnesium is a silvery-white metal with a melting point of 1202°F (1110°C). Pure, sand-cast magnesium has a tensile strength of 12,000 to 13,000 psi. Rolled magnesium has a tensile strength approximately double that of the cast metal—about 25,000 psi. By adding alloying elements, the tensile strength can be considerably increased. For example, the addition of 8 to 10 percent aluminum produces an alloy with a tensile strength of 53,000 psi. A comparison of the properties of magnesium with those of other metals is found in Table 16-1.

Magnesium and magnesium alloys can be forged, machined, and cast. They also can be rolled into plate, shapes, and strips. Because of the high strength to low weight ratio, magnesium and the magnesium alloys have been used as structural metals in aircraft construction.

WELDING MAGNESIUM AND MAGNESIUM ALLOYS

Magnesium and magnesium alloys can be welded by most manual arc welding processes, although gas tungsten arc welding (GTAW) and gas metal arc welding (GMAW) are by far the most commonly used types. Braze welding and brazing are sometimes used to join thin sheets of magnesium and magnesium alloy materials. Soldering has limited use in the repair of surface scratches and dents.

CAUTION

Special safety precautions must be followed when welding or working with magnesium, because the metal is highly oxidizable. For example, the metal turnings and powders produced by machining magnesium can burn with violent intensity when ignited. Always perform machining under strictly controlled conditions and make certain an approved fire extinguishing

Table 16-1 Comparative Properties of Magnesium and Other Metals

| | Approx. Melting Point (°F) | Weight (lb/cu. in.) | Weight (lb/cu. ft.) | Approximate Ratios (Magnesium = 1) | | |
				Weight Ratio	Thermal Conductivity Ratio	Expansion Ratio
Magnesium	1204	0.063	109	1.0	1.0	1.0
Aluminum	1215	0.098	170	1.55	1.4	0.9
Copper	1980	0.323	560	5.1	2.5	0.62
Steel	2700	0.284	490	4.5	0.5	0.45
18–8 Stainless	2600	0.286	495	4.56	0.17	0.63
Nickel	2646	0.322	560	5.1	0.38	0.52

Courtesy The James F. Lincoln Arc Welding Foundation

agent is within reach. Never use water to extinguish a magnesium fire.

Welding magnesium is very similar to welding aluminum, because both share many of the same characteristics. For example, both have relatively low melting temperatures and neither shows a color change as temperatures approaches the melting point. Other important characteristics shared by magnesium and aluminum are high thermal conductivity, a high coefficient of thermal expansion, and excellent corrosion resistance.

NOTE

Magnesium and magnesium alloys require less heat than other metals do in order to reach their melting point. Using too much heat increases the danger of thermal shock and metal distortion.

Magnesium and its alloys should not be welded to other metals, because brittle intermetallic compounds are formed in the process. These compounds prevent the creation of strong welds and joints. Furthermore, attempting to join magnesium and its alloys to other metals increases the possibility of corrosion, which is a factor even when joining two magnesium alloys that differ only slightly in their composition.

Note the following:

- The absence of color change as magnesium approaches its melting temperature is important when using the oxyacetylene welding process, but not when using any of the arc welding processes.

- Magnesium welds do not possess the tensile strength of cold-worked or heat-treated metals; consequently, magnesium alloys are very seldom heat-treated.

- Wrought alloys are usually welded more easily than certain cast alloys.

Gas Tungsten Arc Welding (GTAW)

The gas tungsten arc welding process (GTAW) was originally developed specifically for the welding of magnesium and the magnesium alloys. It is now commonly used for welding thinner sections of these metals, whereas gas metal arc welding (GMAW, or MIG welding) is the preferred welding process for thicker sections.

NOTE

A major factor in preferring the GTAW and GMAW for welding magnesium and the magnesium alloys is the elimination of corrosion problems caused by flux residue after welding by other methods.

Note the Following:

- Use a DC electrode positive (DCEP) or alternating-current welding machine with superimposed high-frequency current when TIG welding magnesium and the magnesium alloys. Never use DC electrode negative (DCEN), because it lacks the proper cleaning action.

- Either alternating current, direct current, or DC electrode positive (DCEP) can be used for welding thin sheets (up to 1/8 inch thick). Alternating current is recommended.

- Work at a uniform travel speed.

- Use only alternating current when welding sheets thicker than 1/8 inch. Alternating current provides deeper penetration.

- Use a short arc with the GTAW torch at a slight leading travel angle.

- Feed in the cold-wire filler metal as near to the horizontal level as possible on flat work.

- Add the filler wire to the leading edge of the weld puddle.

- Use argon as the shielding gas when AC welding on thinner sections. Argon provides better arc stability.

- Add helium to the shielding gas and increase the gas flow rate when welding thicker sections (75-percent helium, 25-percent argon). The addition of helium to the shielding gas will increase arc penetration.

- Use 100-percent helium as the shielding gas on very thick sections.

- Sheets 1/8 to 1/2 inch thick should be beveled to a 90° vee.

- Bevel a double 90° vee along joint edges of sheets thicker than 1/2 inch.

- Remove oxide film on either side of the joint to prevent interference with the inert shielding gas.

- Tack or clamp edges of the sheets and position the work for welding from the top to the bottom, for best results.

- Minimize the possibility of weld cracking by welding from the center toward the sides of the work.

- Use a steel or copper backing plate to avoid overpenetration of the weld metal.

- If a magnesium alloy filler rod is used, it should be of the same composition as the base metal.

- Preheating prior to welding magnesium and its alloys is generally not necessary for thick plates, or when a short weld bead is used.

Gas Metal Arc Welding (GMAW)

Gas metal arc welding is the preferred method for welding thicker magnesium materials because it is faster, which reduces costs. The GMAW process uses three modes of transferring the weld metal through the arc: Short circuit mode, pulsed arc mode, and spray transfer mode (see Table 16-2).

Table 16-2 GMAW Metal Transfer Modes

Transfer Mode	Description	Comments
Short circuit mode	The filler metal touches the work many times per second and extinguishes the arc. The filler metal is supplied in a continuous sequence of drops.	Recommended for use on materials thinner than 3/16 inch
Pulsed arc mode	A special power supply provides a modulated current. The arc is not interrupted, and the filler metal is transferred in a discontinuous, intermediate mode.	Recommended for use on materials 3/16 inch or thicker Most expensive of the transfer modes because of equipment purchase and maintenance costs
Spray transfer mode	The filler metal is transferred as a spray of tiny droplets.	Recommended for use on materials thinner than 3/16 inch.

Note the Following:

- Use a wire feeder with high-speed gear ratios, because the magnesium electrode wire has an extremely high meltoff rate.
- Use a filler metal with the same composition as the base metal when welding magnesium or its alloys. Even slight dissimilarities can cause problems.
- Use argon or mixtures of helium and argon as the shielding gas.

Oxyacetylene Welding

Oxyacetylene welding is sometimes used to weld magnesium and the magnesium alloys when immediate or emergency repairs are required. It works best on materials containing only a small amount of alloying elements.

Note the Following:

- Use a neutral or slightly reducing flame and a salt-powder type flux especially formulated for magnesium.
- Wash any remaining flux off the surface after welding. Flux residue will eventually cause the weld to corrode.
- Limit welding to making butt welds. Lap welds are not recommended because they tend to trap the flux.
- Make allowances for contraction of the metal after welding.

Shielded Metal-Arc Welding (SMAW)

The SMAW process is not generally recommended for welding magnesium alloys. If conditions require it, however, be to thoroughly wash away any flux remaining on the surface after welding. Take particular care to wash the flux out of places where it is likely to be trapped, such as the space between lapped joints. The welds will corrode if any trace of flux remains after welding.

Note the Following:

- Welding magnesium is generally performed by the SMAW process using direct current with reverse polarity (electrode positive).
- Weldability varies with type from excellent to limited.

Braze Welding

Braze welding is a low temperature welding process in which the filler metal is deposited in the joint without melting the base metal. Consequently, expansion and contraction of the metal are much less than found in the fusion welding processes (GTAW, GMAW, SMAW and OAW).

Note the Following:

- Use a single or double V-groove (with a 90° to 120° included angle) for thicknesses greater than 3/32 inch, and a square grove for thicknesses 3/32 inch or thinner.
- Use a suitable braze-welding flux for cleaning the oxides that form on the surface ahead of the torch flame.
- Use a slightly oxidizing torch flame. Adjust the torch to a neutral flame and then open the oxygen needle valve to slightly reduce the length of the inner flame cone.
- Use only approved metal filler rods for braze welding. The Bmg-2a filler metal is generally recommended for torch brazing.
- Clean the surface thoroughly before applying the flux.
- Tin the surface before depositing the braw welding beads in the joint.
- Position the work so that welding can move upward at a slight angle.
- Move the torch flame in a circular motion across a wide area on either side of the surface to avoid a build up of temperatures.
- Preheat a magnesium alloy casting along its weld zone with an oxyacetylene torch. Preheating reduces the amount of oxygen and acetylene required during braze welding.
- Play the torch flame across a wide area around the welded joint after braze-welding to establish a uniform heat level in the metal.
- Cover the surface with a protective material to protect it from cold drafts during after welding. The surface must be allowed to cool evenly.

Brazing

Braze is a low temperature joining process in which the filler metal is distributed between closely fitted surfaces by capillary attraction. Both lap and butt joints can be joined by brazing magnesium and its alloys. Joints should be designed to take advantage of capillary attraction and to allow the flux to be displaced by the brazing filler metal as it flows into the joint. But, because of the corrosive nature of the flux, care should be taken to design the joints for minimal flux entrapment.

Note the following:

- Parts to be brazed should be thoroughly clean and free from burrs. Oil, dirt, and grease should be removed in hot alkaline cleaning baths or by vapor or solvent degreasing.

- Such surface films as chromate conversion coatings or oxides must be removed by mechanical or chemical means immediately prior to brazing.

- When cleaning the surfaces mechanically, abrading with aluminum oxide cloth or steel wool has proved very satisfactory.

- AWS Brazing Fluxes, No. 2 are used to clean the surface of the magnesium, which permits capillary flow. Because of the corrosive nature of these fluxes complete removal is of utmost importance if good corrosion resistance is to be obtained in brazed joints.

- The corrosion resistance of brazed joints depends primarily on the thoroughness of flux removal and the adequacy of joint design to prevent flux entrapment.

- Because of the close proximity of the initial melting point of the base metal (solidus temperature), and the flow point of the brazing filler metal (liquidus temperature, manual torch brazing with BMg-1 filer metal is difficult and requires considerable operator skill.

- Use a BMg-2a filler metal for torch brazing magnesium and magnesium alloys.

- Use tack welding, staking or self-positioning fixtures to assemble and secure the magnesium components before brazing.

- Brazing techniques similar to those used on aluminum are used on magnesium alloys. Torch brazing is accomplished using a neutral flame.

- The filler metal should be placed on the joint and fluxed before heating.

- Avoid overheating the base metal. Overheating can cause the filler metal to penetrate and drop through the bottom of the joint. This is especially a problem with magnesium and magnesium alloys because of their low melting temperatures.

- Thoroughly clean the surface after brazing to remove any flux residue. Hot water and scrubbing with a stiff brush can be used for this purpose.

- Cover the surface with a protective material to protect it from cold drafts during after welding. The surface must be allowed to cool evenly.

Soldering

Magnesium and the magnesium alloys are difficult to solder because of the refractory oxide film that forms on the surface. The film interferes with the flow of the solder. Consequently, soldering is commonly limited to repairing surface defects, such as filling dents or cracks. It should never be used where joints must withstand stress.

Soldering Recommendations:

- Solder will not bond with magnesium or its alloys unless the oxide film is removed.

- Remove the oxide film by mechanical means, such as

cleaning the surface with steel wool or a wire brush. Do not attempt to clean the surface chemically. No effective flux has been developed for this purpose.

- Use a tin, zinc and cadmium solder for magnesium and its alloys.

- A common procedure is to thoroughly clean the surface, pre-coat it with a 70 percent tin, 30 percent zinc solder, and then apply a 60 percent cadmium, 30 percent zinc, 10 percent tin solder.

WELDING MAGNESIUM ALLOY CASTINGS

Welding is commonly used to the repair defects in magnesium alloy castings immediately after being cast or as a result of damage during service. Preparation is most important and should exclude all contamination from extraneous materials.

Note the following:

Make sure you have correctly identified the casting as a magnesium alloy and not an aluminum one.

- Magnesium alloy filings will not clog the file if a scratch test is made (aluminum filings will clog the file).

- Magnesium alloy filings will ignite if heated with an oxy-acetylene flame Aluminum filings turn black.

- Magnesium alloy castings are much lighter in weight than aluminum castings of the same size.

- Magnesium alloy castings are a yellow-gold color when new, but become a dull gray after prolonged use.

- Magnesium alloy castings generally have a lower melting point than aluminum castings.

- Magnesium alloy castings will react with small bubble formations on their surface when tested with an ammonium chloride and water mixture; aluminum castings will show no reaction.

Recommendations for Welding Magnesium Alloy Castings:

- Thoroughly clean and degrease castings before welding.

- Provide wide bevels for full penetration, and use a backing plate to prevent the weld metal from penetrating beyond the bottom of the base metal.

- Heat treatment is generally performed to relieve stress.

- Preheat small areas of magnesium castings with a gas or propane torch, but be careful not to overheat. Temperature change is not indicated by a change in color, as in ferrous metals. Magnesium and magnesium alloys will sag and collapse without any observable warning when a critical temperature is reached.

- Always support the casting to avoid distortions caused by sagging.

- Note: Magnesium alloy castings, unlike the thinner sheet materials, rarely ignite during the welding process. Consequently, they seldom present a safety hazard.

17. LEAD, TIN AND ZINC

Lead, tin, and zinc are soft metals possessing a strong resistance to corrosion, but with low strength and with melting temperatures well below 1000°F.

LEAD AND LEAD ALLOYS

Lead is a grayish-white metal noted for its softness, ductility, and resistance to corrosion. It is also characterized by low strength (it has a particularly low fatigue strength) and a tendency for fatigue cracking. Commercial lead can be as high as 99.99 percent pure. The melting point of pure lead is 621°F (327.5°C).

Lead has many applications, including pipes and other fittings used in plumbing; shielding protection against x-ray and gamma radiation in the dental and medical occupations; corrosion-resistant linings; cable sheathings; and as an alloy with tin in some solders (low-melting types).

Lead is commonly welded by the oxyacetylene welding process. It can also be joined by soldering.

Welding and Joining Tips:

1. Oxyacetylene welding

   a. Welding of lead can be performed in all positions.

   b. The welds are normally stronger than the base metal.

   c. No flux is required, and edge preparation is relatively easy because the metal is very soft.

   d. Use a small torch with small tip sizes.

   e. Use a neutral flame with a slight acetylene excess.

   f. Use a gas pressure less than 5 psi for best results. Note: Ultimately, the gas pressure will depend on the size and type of weld.

   g. Use a filler rod (when required) that matches the

301

analysis of the base metal. A filler rod is not necessary when making flange and edge joints.

h. Pull back the torch when the weld pool comes to the molten stage. Doing so will avoid excess fusion, excess fluidity, and burnthrough, all problems directly related to the low melting point of lead.

i. Other gas welding processes used to weld lead include oxyhydrogen welding and air-acetylene welding. Natural gas in combination with oxygen is also used.

j. The oxyhydrogen welding process is best suited for thinner pieces of lead (generally less than 1/4 inch thick). Thicker pieces of lead are commonly welded with an oxyacetylene torch because of its hotter flame.

2. Soldering

a. Never use solder to join containers, piping, or tubing used to store, carry, or transport corrosive chemicals. These items must be welded.

b. Joint fitup is critical. Maintain close joint tolerances and clamp the sheets or parts in place.

c. Thoroughly clean the surfaces before soldering. A clean surface is necessary to promote capillary attraction—the motive force that draws the solder into the narrow joint.

d. Rosin fluxes are recommended for soldering lead.

e. No post-soldering treatment is required.

TIN AND TIN ALLOYS

Tin has a silvery-white color and a melting point of approximately 449°F (232°C). It is noted for its softness and ductility.

Tin is combined with many other metals to form commonly used metal alloys. For example, tin is mixed with copper to form bronze and with lead to form solder. Tinfoil is produced from a rolled lead-tin alloy. The tin cans used in the food industry are made from thin iron sheets covered on both sides by a coating of tin. Tin is used because it is a nontoxic metal. Pewter is an alloy consisting of at least 91.0 percent tin and small amounts of antimony and copper. These are only a few examples of the many metals in which tin functions as an alloying element. Some of the others include:

- Aluminum-tin alloy (79.0 percent aluminum, 20.0 percent tin, 1.0 percent copper), which is used in the automotive industry for the production of such items as camshaft bearings and connecting rods
- Phosphor bronze (95.0 percent copper, 5.0 percent tin)
- Gunmetal (88.0 percent copper, 10.0 percent tin, 2.0 percent zinc), used for the production of gears and bearings

Note the following:

- Tin alloys are welded with the same techniques used to weld lead alloys.
- Their reaction to heat and other characteristics of the welding process are also similar to those of lead.
- Pure tin is rarely welded.

ZINC AND ZINC ALLOYS

Zinc is obtained from a sulphide ore by an electrolytic process, or through a complicated procedure of separating the zinc from the ore by reducing, boiling, and condensing. Pure zinc has a melting point of 787°F. Its tensile strength is approximately 18,000 psi, but the tensile strength of cast zinc is considerably higher at 40,300 psi.

Zinc is used in its pure form for galvanizing. It is also combined with other metals (aluminum, magnesium, iron, tin) to form zinc castings. Finally, zinc serves as an alloying element in copper-zinc alloys (e.g., zinc bronze).

See Chapter 14 (Galvanized Metals) for information concerning the surface preparation and welding processes for joining galvanized metals.

NOTE

Zinc fumes have a nauseating effect, and they can pose a significant health hazard. These fumes are released in the form of a white vapor at certain temperatures. Several methods can be used for reducing zinc fumes, including the use of an oxidizing flame (when welding with oxyacetylene), special fluxes, and welding at lower temperatures.

18. HARDFACING AND TOOL AND DIE STEELS

Alloy steel has had other metal added to the carbon steel to improve its properties—often, increased strength and hardness. Retention of the metal's properties at elevated temperatures and improvement of the metal's resistance to corrosion are also desirable qualities.

Elements added to steel for alloying are:

- chromium
- nickel
- manganese
- vanadium
- molybdenum
- silicon
- tungsten
- phosphorus
- copper
- titanium
- zirconium
- cobalt
- columbium
- aluminum

These elements form two groups. Those in the first group form a solid solution in which the element dissolves in iron. Silicon, manganese, nickel, molybdenum, vanadium, tungsten and chromium increase the hardness and strength of steel by strengthening the iron base of the alloy. Elements in the second group form complex carbides when cooled slowly. These carbides precipitate out of a solid solution and are scattered throughout the metal. Chromium, manganese, vanadium, tungsten, titanium, columbium, and molybdenum form *carbides* of deep hardening that increase wear resistance and the ability to withstand heat. These alloy carbides can be formed only when the base metal contains sufficient carbon in the metal to form alloy carbides.

HIGH-CARBON STEEL

High-carbon steel has a carbon content ranging from 0.60 to 1.50 percent. It is very strong and hard. Such items as railroad equipment, automobile and truck parts, and farm machinery are made from high-carbon steel. Parts such as springs, grinding balls, bars, hammers, cables, axes, and wrenches are also produced from high-carbon steel with a content of 0.60 to 0.95 percent carbon. This grade of steel has great hardness, toughness, and strength.

When the carbon content is increased from 0.60 to 1.50 percent, the strength and wear resistance increase, but the hardness does not continue to increase. These steels are referred to as tool steels and are used to manufacture cutting and forming tools such as dies, taps, drills, reamers, chisels, and forming and bending dies. The maximum hardness is obtained with 0.80 percent carbon in steel at Rockwell C-66. At 0.60 percent carbon, steel is near maximum hardness.

HARDENING

The process of *hardening* metal is accomplished by increasing the temperature of a metal to its point of *decalescence* and then quenching it in a suitable cooling medium. In actual practice, the temperature of the metal that is to be hardened should be increased to slightly above the point of decalescence for two reasons:

- To be certain that the temperature of the metal is above the point of decalescence at all times

- To allow for a slight loss of heat while transferring the metal from the furnace to the quenching bath

Steel with less than 0.85 percent carbon needs to be heated above its upper critical temperature.

Heating Process

Steel can be heated most successfully by placing it in a cold furnace and then bringing the furnace and its charge to the hardening temperature slowly and uniformly. Commercially, only those steels that are difficult to harden are heat-treated in this. For example, some tool steels are heated for hardening by first placing them in a preheating furnace at a temperature of 1000°F (537.77°C) or slightly lower. After the steel is heated uniformly to furnace temperature, it is withdrawn and placed in the high-heat furnace at the hardening temperature. When the steel has become heated uniformly at the hardening temperature, it is then quenched to harden it.

During the heating process, the metal absorbs heat and its temperature rises, until the point of decalescence is reached. At this point, the metal takes up additional heat. This heat is converted into work that changes the pearlite into austenite without an increase in temperature, until the process is completed.

This phenomenon can be compared to the *latent heat of steam* in that when the temperature of water is increased to the boiling point, it will absorb an additional quantity of heat without an increase in temperature. The heat is converted into work that is necessary:

- To bring about a change of state from a liquid to a gas
- To overcome the pressure of the atmosphere in making room for the steam

The total latent heat required to bring about these changes consists of *internal* latent heat and *external* latent heat.

After the metal has been heated, it is *quenched* to permanently fix the structural change in the metal. Quenching causes the metal to remain hard after it has been heated to the point of decalescence. If the metal were not quenched and allowed to cool slowly, the austenite would be reconverted to pearlite as the temperature decreased. That would cause the metal to lose its hardness. When steel is cooled faster than its critical cooling rate, which is the purpose of quenching, a new structure is formed. The austenite is transformed into *martensite*, which has an angular, needle-like structure and a very high degree of hardness.

Martensite has a lower density than austenite, so the steel will increase in volume when quenched. Some of the austenite will not be transformed during quenching, but it will gradually change into martensite over a period of time. This change is known as *aging*. This aging results in an increased volume that is objectionable in many items, such as gauges. The cold-treating process can be used to eliminate this problem.

Metal bars will decrease in length when quenched in water at a temperature below the critical point. If a metal bar is quenched in water at a temperature above the critical point, hardening also will be indicated.

Pyrometers are used in production work to indicate the critical temperature, if the critical temperature of a given steel is known. A magnetic needle can also be used to determine whether steel has been heated above the critical point. When heated above the critical point, a piece of steel loses its magnetism. However, it will attract a magnetic needle if it has been heated to any temperature below the critical point. In making the test with a magnetic needle, the magnetic influence of the cold tongs should be eliminated. A pivoted bar magnet can be introduced into the furnace momentarily to determine the presence or absence of magnetism in the piece of steel.

Heating Baths
Other methods of heating are used in the hardening process, including various liquid baths: (1) lead, (2) cyanide of potassium, (3) barium chloride, and (4) a mixture of barium chloride and potassium chloride, or other metallic salts.

The chief advantage of the liquid bath is that it helps to prevent overheating of the workpiece. The work cannot be heated to a temperature higher than the temperature of the bath in which it is heated. Other advantages of the liquid bath are:

- The temperature can be easily maintained at the desired degree level.
- The submerged steel can be heated uniformly.
- The finished surfaces are protected against oxidation.

The lead bath cannot be used for high-speed steels because it begins to vaporize at 1190°F (643.33°C). The cyanide of potassium bath is used in gun shops extensively to achieve ornamental effects and to harden certain parts.

A thermoelectric pyrometer can be used to indicate the temperature of a barium chloride bath. Potassium chloride can be added for heating the bath to various temperatures:

- For temperatures from 1400°F (760°C) to 1650°F (899°C), use three parts barium chloride to one part potassium chloride.
- For higher temperatures, reduce the proportion of potassium chloride.

Quenching or Cooling baths

Although water is one of the poorest conductors of heat among liquids, it is the most commonly used liquid for quenching metals in heat-treating operations. Water cools by means of evaporation.

To raise the temperature of one pound of water from 32°F (0°C) to 212°F (100°C) requires 180 Btu; converting it into steam requires an additional 950 Btu. Thus, the latent heat of vaporization (950 Btu) absorbed from the hot metal is the cooling agency. For efficient cooling, the film of steam must be replaced immediately by another layer of water—and this requires circulation. Plunging the heated metal into the water causes thermocirculation. However, if the metal is placed in the water horizontally, the film of steam that forms on the lower side of the piece is *pocketed*, and the cooling action is greatly retarded. In still-bath quenching, a slow up-and-down movement of the work is recommended.

In addition to plain water, several other solutions can be used for quenching. Salt water, water and soap, mercury, carbonate of lime, wax, and tallow can also be used. Salt water will produce a harder scale. The quenching medium that will cool rapidly at the higher temperatures and more slowly at the lower temperatures should be selected. Oil quenches meet this requirement for many types of steel. There are various kinds of oils employed, depending on the nature of the steel being used.

The quenching bath should be kept at a uniform temperature so that successive pieces quenched will be subjected to the same conditions. The next requirement is to keep the bath agitated. The volume of work, the size of the tank, the method of circulation, and even the method of cooling the bath need to be considered in order to produce uniform results.

High-speed steel is usually quenched in oil, but a blast of air under pressure is used to quench many high-speed steels. Air under pressure is applied to the work, that is, by air blast.

TEMPERING

The purpose of *tempering*, or drawing, is to reduce brittleness and remove internal strains caused by quenching.

The process of tempering metal is accomplished by reheating steel that has been hardened previously, and then quenching it to toughen the metal and make it less brittle. Unfortunately, the tempering process also softens the metal.

Tempering is a reheating process for which the term hardening is often used erroneously. In the tempering process, the metal is heated to a much lower temperature than is required for hardening. Reheating to a temperature between 300°F (149°C) and 750°F (399°C) causes the martensite to change to *troostite*, a softer but tougher structure. Reheating

Table 18-1 Typical Tempering Temperatures for Certain Tools

Degrees Fahrenheit	Degrees Celsius	Temper Color	Tools
380	193	Very light yellow	Lathe centers, cutting tools for lathes, and shapers
425	218	Light straw	Milling cutters, drills, and reamers
465	241	Dark straw	Taps, threading dies, punches, hacksaw blades
490	254	Yellowish brown	Hammer faces, shear blades, rivet sets, and wood chisels
525	274	Purple	Center punches and scratch awls
545	285	Violet	Cold chisels, knives, and axes
590	310	Pale blue	Screwdrivers, wrenches, and hammers

to a temperature between 750°F (149°C) and 1290°F (699°C) causes a structure known as *sorbite* to be formed. It has less strength than troostite, but much greater ductility. See Table 18-1 for the temperatures required for tempering various tools.

Color Indications
Steel that is being heated becomes covered with a thin film of oxidation that grows thicker and changes in color as the temperature rises. This variation in color can be used as an indication of the temperature of the steel and the corresponding temper of the metal.

As the steel is heated, the film of oxides passes from a pale yellow color, through brown, to blue and purple colors. When the desired color appears, the steel is quenched in cold water or brine. Using a microscope can explain the phenomena associated with the change in color. Steel consists of various manifestations of the same compound rather than separate compounds.

Although the color scale of temperatures has been used for many years, it gives only rough or approximate indications that vary for different steels. The color scale for temper colors and corresponding temperatures is given in Table 18-2.

Table 18-2   Temper Colors of Steel

| | Temperatures | |
Colors	(Fahrenheit)	(Celsius)
Very pale yellow	430	221
Light yellow	440	227
Pale straw-yellow	450	232
Straw-yellow	460	238
Deep straw-yellow	470	243
Dark yellow	480	249
Yellow-brown	490	254
Brown-yellow	500	260
Spotted red-brown	510	266
Brown-purple	520	271
Light purple	530	277
Full purple	540	282
Dark purple	550	288
Full blue	560	293
Dark blue	570	299
Very dark blue	600	316

Specially prepared tempering baths equipped with thermometers provide a more accurate method of tempering metals. They are much more accurate than the color scale for tempering.

CASEHARDENING

The casehardening operation is a localized process in which a hard skin, or surface, is formed on the metal to a depth of 1/16 (0.0625) to 3/8 (0.375) inch. Applying this hard surface, or *case*, requires two operations: (1) carburizing the outer surface to impregnate it with sufficient carbon, and (2) heat-treating the carburized parts to obtain a hard outer case while giving the core of the metal the required physical properties. Casehardening usually refers to both operations together.

Carburizing is accomplished by heating the work to a temperature below its melting point in the presence of a material that liberates carbon at the temperature used. The material can be solid (charcoal, coke, etc.), liquid (sodium cyanide, other salt baths), or gas (methane, propane, butane). Often only part of the work is to be casehardened. In that case, there are four distinct ways to eliminate casehardening from portions of the work:

1. Copper plating
2. Covering the portion that is not to be hardened with fire clay
3. Using a bushing or collar to cover the portion that is to remain soft
4. Packing in sand

An article that is to be casehardened can be copper-plated on the portion that is to remain soft. This is especially useful when a liquid carburizing process is used. The portion to remain soft can also be protected by covering with fire clay, covering with a collar or bushing, or packing the portion in sand. The size and shape of the work, will dictate the methods as well as the type of steel, and the process.

The casehardening furnace must provide a uniform heat, and steel that is to be casehardened must be selected carefully. Since oil and gas have superseded coal as fuels for casehardening furnaces, furnace construction has changed considerably. Careful consideration should be given to both the carbonaceous material used in packing the parts and the box in which the material is packed. The operation of packing the insulated workpiece is referred to as preparation for: *local hardening.*

In the casehardening process, the metal items packed with carbonaceous material are heated to a cherry red color in a closed vessel, and then quenched suddenly in a cooling bath. Malleable castings can be casehardened so that they acquire a polish. Malleable iron can be casehardened by heating it to a red heat, rubbing cyanide of potassium over the surface or immersing in melted cyanide, reheating, and then quenching the piece in water. Hardening is usually a separate operation following the carburizing and is designed for the part and the steel used.

Both iron and steel can be casehardened, but it is used mostly on steel products. For example, the gears of automobile transmissions are casehardened so that they can withstand the abuse of shifting gears (not synchromesh) without waiting for them to synchronize.

Variations on Casehardening Methods

A commonly used method of casehardening is to carburize the material and then allow the boxes to cool with the work in them, after which they are reheated and then hardened in water. It is satisfactory to dump items such as bolts, nuts, and screws into the water directly from the carburizing furnace, without reheating.

A common iron wheelbarrow with two pieces of flat iron placed across it lengthwise should be provided. A sieve made of 1/8-inch wire woven in a 1/4-inch mesh and approximately 18 inches square by 6 inches deep is placed on the bars. The sieve should have a handle 5/8 inch in diameter and 6 feet long. The carburizing boxes are emptied into the sieve and the carbonaceous material is sifted off through the mesh. The hot metal items are then dumped into a tank of cold water, which should be large enough to prevent the water from heating too quickly. Care should be taken not to empty the entire contents of the boxes into the water in one place, and a constant flow of water should be available while the work is being hardened.

The work should never be removed from the furnace until the temperature has been lowered, because it is harmful to the steel to harden it at the high carburizing temperature. The steel should be treated as tool steel after it is carburized.

Gears and other parts should be tough, but not extremely hard. They should be hardened in an oil bath to make the work less liable to warping and enable the hardened product to withstand shocks and severe stresses without breakage.

Annealing is a process of softening metal by heating it to a high temperature and then cooling it very slowly. The objective of the annealing processes is to remove the stresses and strains set up in the metal by rolling or hammering, so that it will be soft enough for machining.

The process of hardening is accomplished by increasing the temperature of a metal to the point of decalescence and then quenching it in a suitable cooling medium. The process of tempering is accomplished by reheating steel that has been hardened previously and then quenched to toughen the metal and make it less brittle. Tempering processes also can soften certain metals.

HARDFACING

Hardsurfacing, or *hardfacing*, is the melting of very hard metal powders and welding them onto the surface of the base metal. The welder can also use a special type of welding rod to hardsurface a tool.

Some tools and machinery are subjected to much abuse. For example, the farmer has rocks in the field, so a plow blade takes a lot of punishment and needs attention very often. The blade on a bulldozer also wears quickly and needs hardening.

Tools can be protected against wear by hardening or coating the edge of the tool that receives the punishment. One method used to do this is heating the forward area with a welding torch and then melting a special metal over it. The additional new metal will be extremely resistant to wear and will help prolong the life of the tool.

Among the most common hardsurfacing materials are nonferrous cobalt-chromium-tungsten alloys. These include nickel-base, iron-base, and tungsten carbide materials. Local welding rod supply dealers have information on each of the rods and where it can be used, and the supply house usually has a catalog that lists all the rods and specialties.

TOOL AND DIE STEELS

Tool and die steels are classified as:

- High-speed
- Hot-work
- Cold-work
- Shock-resisting
- Mold steels
- Special-purpose
- Water-hardening

These basic types of tool and die steels are described in Table 18-3.

In most instances, a good grade of tool steel is used for making punches and dies. The steel should be free from harmful impurities. Sometimes the body of the die can be made of cast iron, with an inserted steel bushing to reduce the cost of material. An advantage of this type of construction is that the insert can be replaced when it becomes worn. Soft steel that has been casehardened does not change its form as readily as tool steel, and any minute changes in form can be corrected readily because the interior is soft.

Table 18-3 Tool and Die Steels

TYPE	AISI	
High-speed	M	(molybdenum base)
	T	(tungsten base)
Hot-work	H1 to H19	(chromium base)
	H20 to H39	(tungsten base)
	H40 to H59	(molybdenum base)
Cold-work	D	(high-carbon, high-chromium)
	A	(medium-alloy, air-hardening)
Shock-resisting	O	(oil-hardening)
	S	
Mold steels	P1 to P19	(low-carbon)
	P20 to P39	(others)
Special-purpose	L	(low-alloy)
	F	(carbon-tungsten)
Water-hardening	W	

Interior strains or stresses are set up in steel during the manufacturing process. In die-making operations, these stresses must be relieved before the die is brought to its final size or they will cause distortion. The presence of stresses cannot be determined in the steel beforehand, but the diemaker can relieve stresses in the steel by annealing it after the die has been roughed out.

19. TYPES OF JOINTS

There are five basic types of joints: (1) the butt joint, (2) the corner joint, (3) the edge joint, (4) the lap joint, and (5) the tee-joint (See Figure 19-1). These basic joints may be used in combination to produce a number of different variations.

BUTT JOINT

The *butt joint* is a joint formed between two members that lie approximately on the same plane. This type of joint is formed when two plates or pipes are brought together, edge to edge and welded along the seam formed between the two members and the ends of the pipe.

CORNER JOINT

The *corner joint* is a joint formed between two members that lie at approximately a right angle to each other.

EDGE JOINT

The *edge joint* is a joint formed between the edges of two or more members that lie parallel or nearly parallel.

LAP JOINT

The *lap joint* is a joint formed by two overlapping members. The welding material is so applied as to bind the edge of one plate to the face of the other. This is to allow for the contraction stresses set up as the weld progresses.

TEE JOINT

The *tee joint* is a joint formed between two members that are located approximately at right angles to each other. The resulting connection forms a tee.

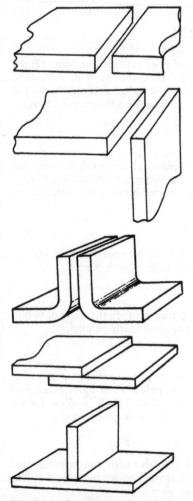

Fig. 19-1 Basic types of joints. *(Courtesy American Welding Society)*

20. WELDING POSITIONS

The *welding position* is the position used to perform the welding and deposit the filler metal in the joint. The four basic welding positions are: (1) the flat position, (2) the horizontal position, (3) the vertical position, and (4) the overhead position.

FLAT POSITION

Flat-position welding (see Figure 20-1) is performed from the upper side of the joint, and the face of the weld is approximately horizontal. This is the most commonly used welding position, because the molten weld metal is not affected by gravity as it is in the other welding positions. As a result, it is easier to deposit a uniform bead with proper penetration at a fast rate. Point the tip of the torch or the electrode downward when welding in the flat position.

HORIZONTAL POSITION

Horizontal-position welding (see Figure 20-2) has two basic forms, depending on whether it is used for a fillet weld or a groove weld. In a horizontal fillet weld, the welding is performed on the upper side of an approximately horizontal surface and against an approximately vertical surface. In a groove weld, the axis of the weld for the horizontal plane and the face of the weld lie in an approximately vertical plane. In both cases, gravity can cause the molten weld metal to sag before a uniform bead with sufficient penetration can be deposited. Point the tip of the torch or the electrode backward toward the weld pool when welding in the horizontal position.

VERTICAL POSITION

In *vertical-position* welding (see Figure 20-3) the weld material is applied to a vertical surface, or one inclined 45° or less to the vertical. On plates, the direction of welding can

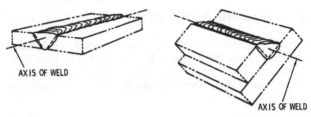

Fig. 20-1 Flat position for a groove weld (left) and a fillet weld (right).

be either from the bottom to the top or from the top to the bottom. If you are welding in an upward direction in the joint, point the tip of the torch or electrode at an angle ahead of the molten weld pool. If, on the other hand, the direction of travel is downward in the joint, point the tip of the torch or electrode up and at an angle to the weld pool.

OVERHEAD POSITION
Overhead-position welding (see Figure 20-4) is performed from the underside of the joint. This welding position requires considerable skill and experience to lay a uniform bead with

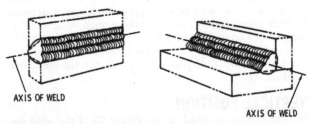

Fig. 20-2 Horizontal position for a groove weld (left) and a fillet weld (right).

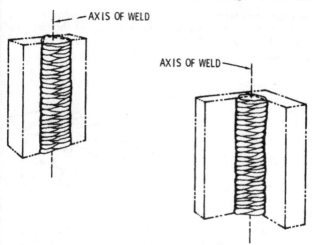

Fig. 20-3 Vertical position for a groove weld (left) and a fillet weld (right).

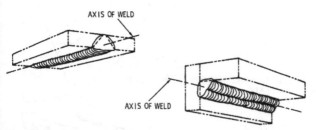

Fig. 20-4 Overhead position for a groove weld (left) and a fillet (right) weld.

sufficient penetration. The weld metal has a tendency to sag and pull away from the joint. Point the tip of the torch or electrode upward in the joint when welding in the overhead position.

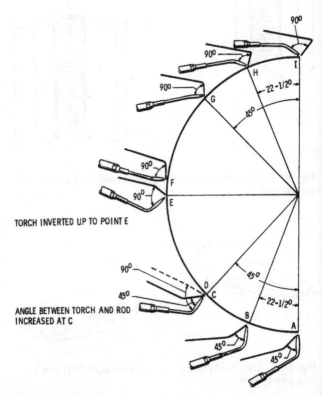

Fig. 20-5 Various torch tip and filler rod angles for welding a pipe in a fixed position.

FIXED-POSITION WELDING

The welder will sometimes be required to weld pipe that must remain stationary and cannot be rolled during welding. Either vertical or horizontal welding in a fixed position must be used, depending on whether the pipe is in the vertical or horizontal position. In either case, the angle of the torch tip and filler rod (oxyacetylene welding) or electrode (arc welding) must be constantly changed to adjust to the curvature of the pipe. Figure 20-5 illustrates the changing angle of the torch tip and filler rod when gas welding a pipe in a fixed position.

21. TYPES OF WELDS

The two most common types of welds are the *groove* weld and *fillet* weld. Both types have a number of variations, depending on joint design. Others types of welds commonly used include the *flange* weld, *surfacing* weld, *plug* weld, *slot* weld and *tack* weld.

GROOVE WELD

The *groove weld* is made in a groove between two adjacent surfaces in the horizontal plane. Groove welds require edge preparation of the joining surfaces. Various types of groove welds are illustrated in Figure 21-1.

FILLET WELD

The *fillet weld* is a weld joining two surfaces positioned at right angles to one another in a lap joint, a tee-joint, or a corner joint. The cross section of a fillet weld is approximately triangular, with either a convex or a concave face (see Figures 21-2 and 21-3).

A *full* fillet weld is one whose size, specifically, "width," is equal to the thickness of the thinner of the two members being joined.

Staggered intermittent fillet welds (see Figure 21-4) consist of two lines of intermittent fillet welds on a joint. The weld increments in one line are staggered with respect to those in the other. *Chain intermittent* welds (see Figure 21-5), on the other hand, consist of two lines of intermittent fillet welds in which the weld increments are located opposite one another.

FLANGE WELD

A *flange* weld is made on the edges of two or more members to be joined when at least one of them has been flanged (turned back at a right angle to the surface for reinforcement). The *corner* flange weld (see Figure 21-6) is an example in which the edge of only one member has been flanged. In the *edge*

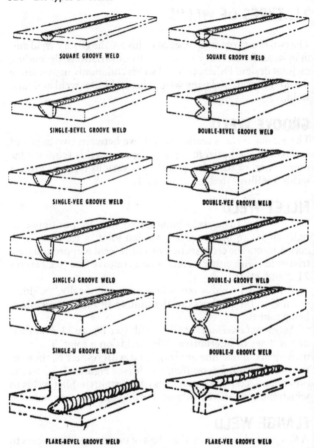

SQUARE GROOVE WELD

SQUARE GROOVE WELD

SINGLE-BEVEL GROOVE WELD

DOUBLE-BEVEL GROOVE WELD

SINGLE-VEE GROOVE WELD

DOUBLE-VEE GROOVE WELD

SINGLE-J GROOVE WELD

DOUBLE-J GROOVE WELD

SINGLE-U GROOVE WELD

DOUBLE-U GROOVE WELD

FLARE-BEVEL GROOVE WELD

FLARE-VEE GROOVE WELD

Fig. 21-1 Types of groove welds. *(Courtesy American Welding Society)*

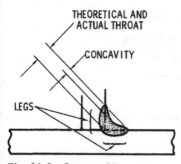

Fig. 21-2 Concave fillet weld.

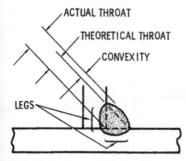

Fig. 21-3 Convex fillet weld.

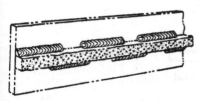

Fig. 21-4 Staggered intermittent fillet weld.

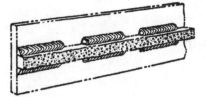

Fig. 21-5 Chain intermittent weld.

Fig. 21-6 Corner flange weld.

Fig. 21-7 Edge flange weld.

flange weld (see Figure 21-7), the edges of both members have been flanged. Flange welds are frequently used in sheet-metal work, and filler metals are not necessary.

SURFACING WELD

A *surfacing* weld is deposited on an unbroken surface in order to produce certain properties or dimensions (see Figure 21-8). These welds consist of one or more string or weave beads. A surfacing weld is sometimes referred to as a *bead* weld.

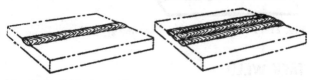

Fig. 21-8 Surfacing weld.

PLUG WELD

A *plug* weld (see Figure 21-9) is a circular weld made through a hole in one member of a lap or tee joint in order to join that member to the other.

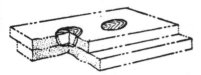

Fig. 21-9 Plug weld.

SLOT WELD

A *slot* weld (See Figure 21-10) is also used with the lap or tee joint, but it is distinguished from the plug weld by a more elongated hole to weld through.

Fig. 21-10 Slot weld.

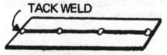

Fig. 21-11 Tack weld. *(Courtesy Airco Welding Products)*

TACK WELD

A *tack* weld is a temporary weld used to hold two parts in position until a more permanent weld can be made (see Figure 21-11). It is not intended for use as a permanent weld.

22. WELD TERMINOLOGY

There are certain basic terms used to describe the structure of a weld and its relationship to adjacent joint surfaces. Most of the terms defined in this chapter apply to groove and fillet welds, the two most commonly used types of welds.

ROOT OF THE WELD

The *root of the weld* (see Figure 22-1) refers to the points at which the bottom of the weld intersects the surfaces of the base metal. In a fillet weld, it is specifically the point of deepest penetration of the weld material.

ROOT FACE, GROOVE FACE, AND ROOT EDGE

Root face, *groove face*, and *root edge* (see Figure 22-2) are terms describing the surfaces in the groove cavity or along the edges that will touch when the two members are joined. The root face in a groove weld represents that portion of the groove face adjacent to the root of the joint. In the butt joint and tee joint, the groove face and the root face are represented by the same surface. The groove face, then, is that surface portion of a member that is included in the groove. The root edge is a root face with zero width.

ROOT OPENING, BEVEL ANGLE, AND GROOVE ANGLE

Root opening, *bevel angle*, and *groove angle* (see Figure 22-3) are all terms referring to space dimensions between the two members to be joined. The root opening is the amount of separation between the two members at the root of the joint. The bevel angle is the angle formed between a plane perpendicular to the surface of the member and its prepared edge. The groove angle, on the other hand, includes the total angle of the groove between members to be joined by a groove weld.

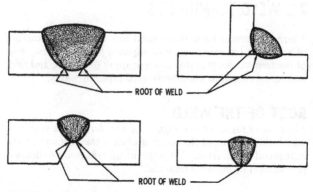

Fig. 22-1 The root of a weld.

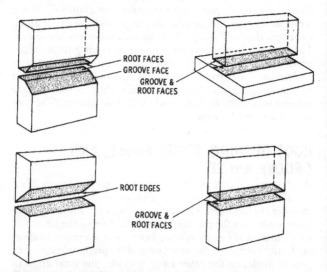

Fig. 22-2 Root face, groove face, and root edge.

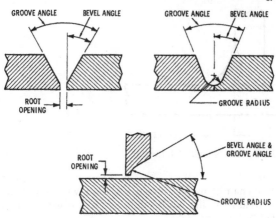

Fig. 22-3 Root opening, bevel angle, and groove angle.

SIZE OF WELD

The meaning of the term *size of weld* depends on whether a groove weld or a fillet weld is being described. In a groove weld (see Figure 22-4), the term indicates joint penetration

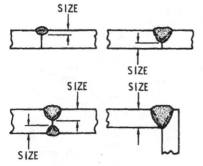

Fig. 22-4 Size of weld (groove weld).

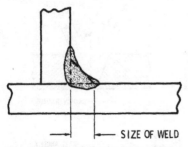

Fig. 22-5 Size of weld (fillet weld).

(depth of chamfering plus root penetration, when specified). Size of weld in a fillet weld is indicated by the leg length of the fillet (see Figure 22-5).

FACE OF A WELD

The *face* of a weld (See Figure 22-6) is the exposed surface of a weld on the side from which the welding is done.

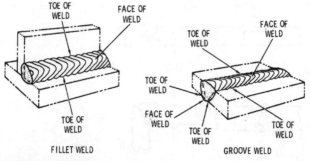

Fig. 22-6 The face of a weld.

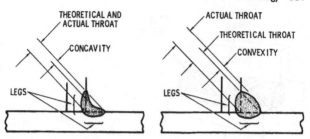

Fig. 22-7 Concave and convex fillet welds, illustrating both actual and theoretical throats.

THROAT OF A FILLET WELD

The *throat* of a fillet weld can be both theoretical and actual (See Figure 22-7). The *actual throat* is the shortest distance from the root of the fillet weld to its face. After the largest possible right triangle is inscribed within the fillet weld's cross section, the *theoretical throat* is the distance from the beginning of the root to the hypotenuse of that triangle.

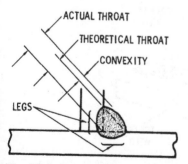

Fig. 22-8 Leg of a fillet weld.

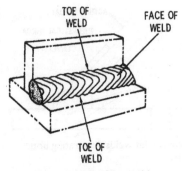

Fig. 22-9 Toe of a fillet weld.

LEG OF A FILLET WELD

The *leg* of a fillet weld (See Figure 22-8) represents the distance from the root of the joint to the toe of the fillet weld.

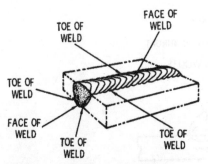

Fig. 22-10 Toe of a groove weld.

TOE OF A FILLET WELD

The *toe* of a weld (see Figures 22-9 and 22-10) is the junction between the face of the weld and the base (parent) metal.

DEPTH OF FUSION

The distance that the *fusion* (melting of the metal) extends into the base metal.

23. WELDING SYMBOLS

The American Welding Society has developed a standardized set of symbols and definitions for the various welding and cutting processes. Always consult the latest AWS publication for current usage of welding symbols and definitions, because it will include any additions, modifications, or deletions of material will. These changes, which are made regularly, are the responsibility of the AWS Committee on Definitions and Symbols.

Welding symbols provide the welder with the following important items of information:

- Type of joint
- Type of weld
- Size of weld
- Amount of deposited (filler) metal
- Location of the weld

A typical welding symbol consists of the following eight elements:

- Reference line
- Arrow
- Basic weld symbols
- Dimensions and other data
- Supplementary symbols
- Finish symbols
- Tail
- Specifications, process, or other references

The standard location of these elements in a welding symbol is illustrated in Figure 23-1. In short, a welding symbol is an assembly of elements that conveys certain specific welding

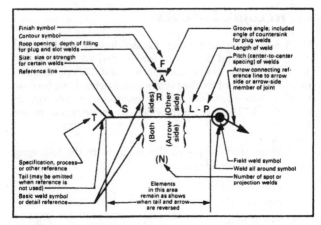

Fig. 23-1 Standard locations of elements in a welding symbol.
Courtesy American Welding Society

information to the welder. Some typical welding symbols are shown in Figure 23-2.

Note that a distinction is made between the terms *weld symbol* and *welding symbol*. A *weld symbol* is an ideograph used to indicate the type of weld to be used. The basic types of weld symbols are shown in Figure 23-3. There are also supplementary weld symbols that are used with the basic weld symbols to provide additional information (Figures 23-4 and 23-5). Both the basic and the supplementary weld symbols represent only two of the eight elements of a *welding* symbol.

The open tail of the arrow in a welding symbol often contains a letter designating the welding or cutting process to be used in making the weld. It may also contain welding specifications, procedures, or other supplementary information for making the weld. If this information is already known, the notations do not have to be used, and the tail of the welding symbol may be omitted.

The joint side in contact with the arrow point of the welding symbol is referred to as the *arrow side* of the joint. The side opposite the arrow side of the joint is referred to as the *other side*. Each joint indicated by a welding symbol will always have an *arrow side* and an *other side* so as to locate the weld with respect to the joint.

Figures 23-6 through 23-44 represent the applications of the different types of welding symbols.

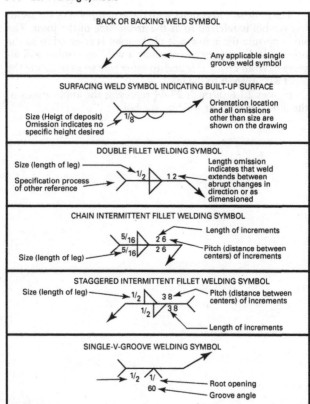

Fig. 23-2 Some typical welding symbols. *Courtesy American Welding Society*

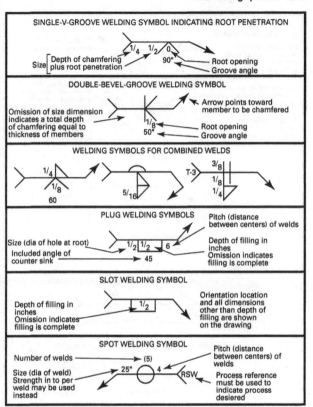

Fig. 23-2 *(continued)*

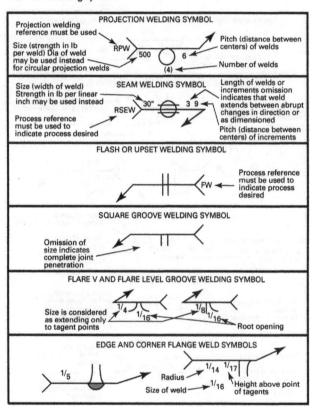

PROJECTION WELDING SYMBOL

Projection welding reference must be used

Size (strength in lb per weld) Dia of weld may be used instead for circular projection welds

RPW   500   6

Pitch (distance between centers) of welds

(4)

Number of welds

SEAM WELDING SYMBOL

Size (width of weld) Strength in lb per linear inch may be used instead

Process reference must be used to indicate process desired

RSEW   30"   3 9

Length of welds or increments omission indicates that weld extends between abrupt changes in direction or as dimensioned

Pitch (distance between centers) of increments

FLASH OR UPSET WELDING SYMBOL

FW

Process reference must be used to indicate process desired

SQUARE GROOVE WELDING SYMBOL

Omission of size indicates complete joint penetration

FLARE V AND FLARE LEVEL GROOVE WELDING SYMBOL

Size is considered as extending only to tagent points

1/4   1/16

1/8   1/16

Root opening

EDGE AND CORNER FLANGE WELD SYMBOLS

1/5

1/14   1/17

Radius

Size of weld   1/16

Height above point of tagents

Fig. 23-2 (*continued*)

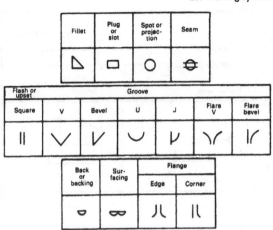

Fig. 23-3 Basic weld symbols. *Courtesy American Welding Society*

Weld all around	Field weld	Melt-thru	Contour		
			Flush	Convex	Concave

Fig. 23-4 Supplementary weld symbols. *Courtesy American Welding Society*

Fig. 23-5 Supplementary symbols used with welding symbols. *Courtesy American Welding Society*

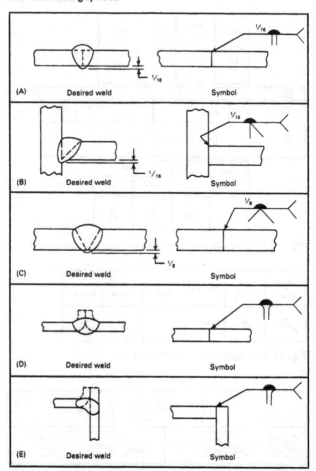

Fig. 23-6 Application of melt-through symbol. *Courtesy American Welding Society*

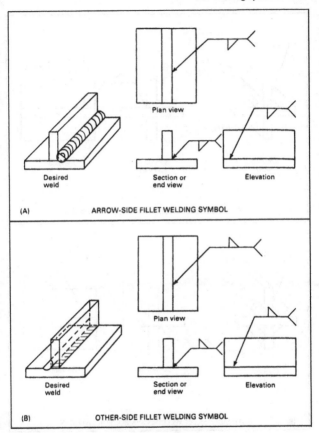

Fig. 23-7 Application of fillet welding symbol. *Courtesy American Welding Society*

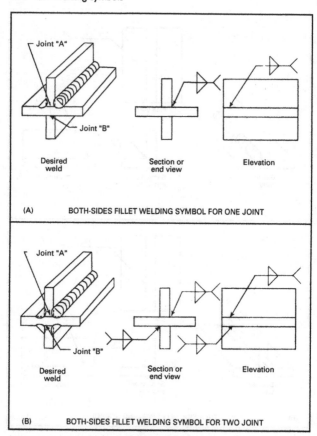

Fig. 23-8 Application of fillet welding symbol.
Courtesy American Welding Society

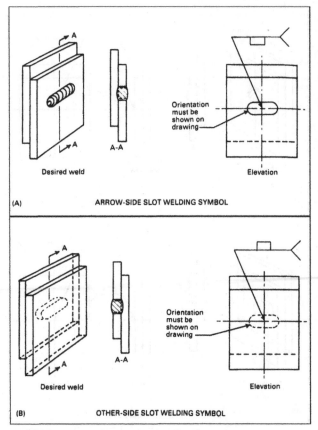

Fig. 23-9 Application of plug welding symbol. *Courtesy American Welding Society*

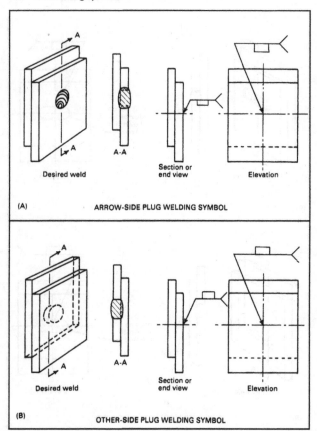

(A)   ARROW-SIDE PLUG WELDING SYMBOL

Desired weld   A-A   Section or end view   Elevation

(B)   OTHER-SIDE PLUG WELDING SYMBOL

Fig. 23-10 Application of slot welding symbol. *Courtesy American Welding Society*

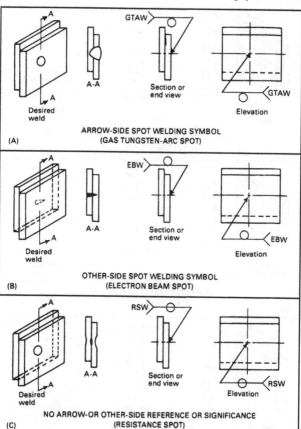

Fig. 23-11 Application of spot welding symbol. *Courtesy American Welding Society*

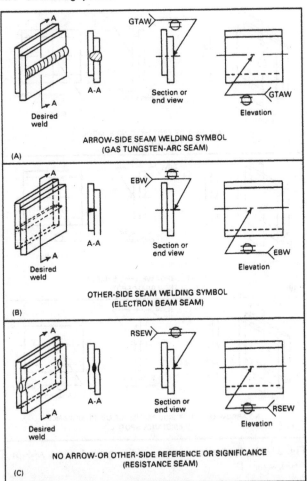

Fig. 23-12 Application of seam welding symbol. *Courtesy American Welding Society*

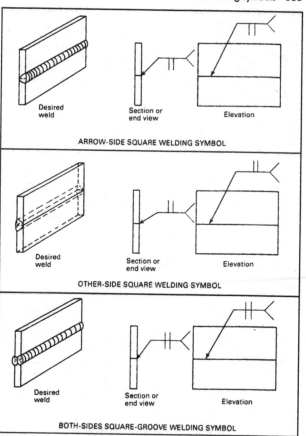

Fig. 23-13 Application of square-groove welding symbol.
Courtesy American Welding Society

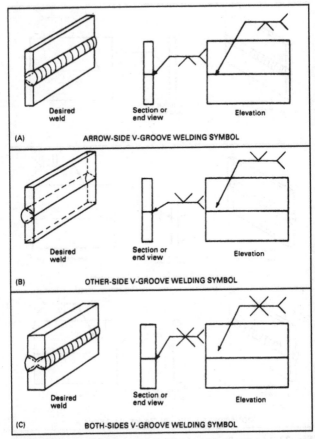

Fig. 23-14 Application of V-groove welding symbol. *Courtesy American Welding Society*

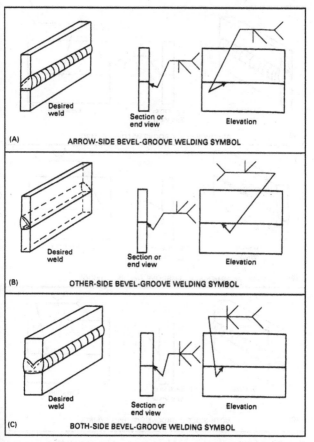

Fig. 23-15 Application of bevel-groove welding symbol.
Courtesy American Welding Society

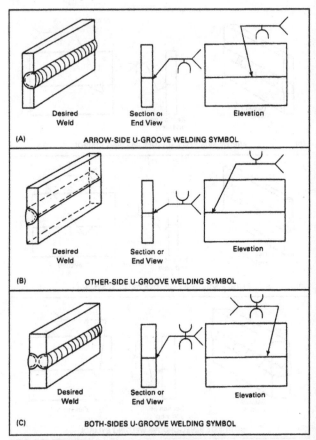

(A) ARROW-SIDE U-GROOVE WELDING SYMBOL

Desired Weld

Section or End View

Elevation

(B) OTHER-SIDE U-GROOVE WELDING SYMBOL

Desired Weld

Section or End View

Elevation

(C) BOTH-SIDES U-GROOVE WELDING SYMBOL

Desired Weld

Section or End View

Elevation

Fig. 23-16 Application of U-groove welding symbol. *Courtesy American Welding Society*

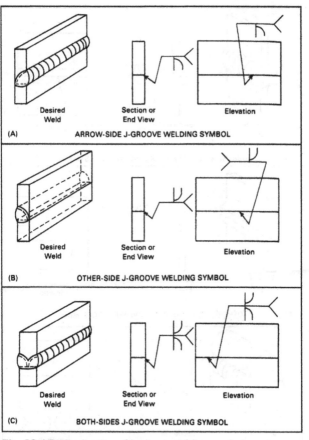

Desired Weld **Section or End View** **Elevation**

(A)   ARROW-SIDE J-GROOVE WELDING SYMBOL

Desired Weld **Section or End View** **Elevation**

(B)   OTHER-SIDE J-GROOVE WELDING SYMBOL

Desired Weld **Section or End View** **Elevation**

(C)   BOTH-SIDES J-GROOVE WELDING SYMBOL

Fig. 23-17 Application of J-groove welding symbol.
Courtesy American Welding Society

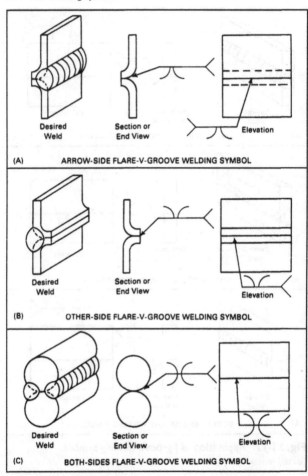

Fig. 23-18 Application of flare-V-groove welding symbol.
Courtesy American Welding Society

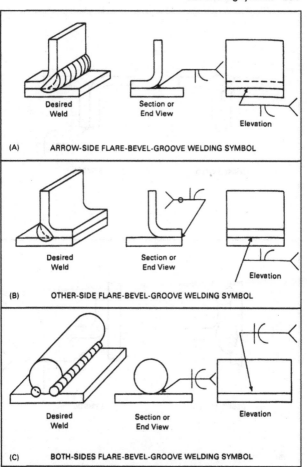

(A) ARROW-SIDE FLARE-BEVEL-GROOVE WELDING SYMBOL

Desired Weld

Section or End View

Elevation

(B) OTHER-SIDE FLARE-BEVEL-GROOVE WELDING SYMBOL

Desired Weld

Section or End View

Elevation

(C) BOTH-SIDES FLARE-BEVEL-GROOVE WELDING SYMBOL

Desired Weld

Section or End View

Elevation

Fig. 23-19 Application of flare-bevel-groove welding symbol.
Courtesy American Welding Society

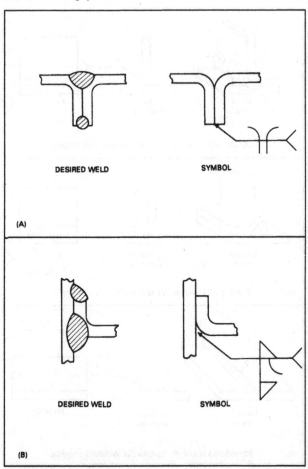

(A)

DESIRED WELD   SYMBOL

(B)

DESIRED WELD   SYMBOL

Fig. 23-20   Application of flare-bevel and flare-V-groove welding symbols. *Courtesy American Welding Society*

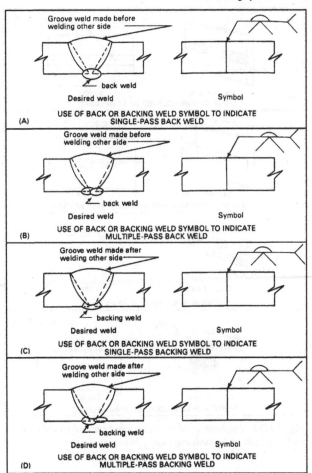

Fig. 23-21 Application of back or backing weld symbol. *Courtesy American Welding Society*

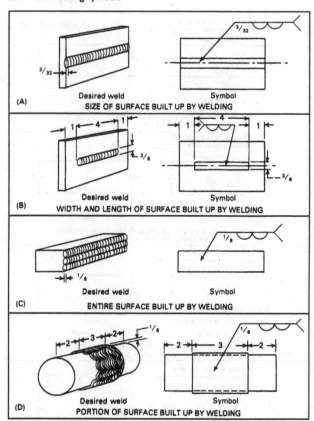

(A) SIZE OF SURFACE BUILT UP BY WELDING

Desired weld　　Symbol

(B) WIDTH AND LENGTH OF SURFACE BUILT UP BY WELDING

Desired weld　　Symbol

(C) ENTIRE SURFACE BUILT UP BY WELDING

Desired weld　　Symbol

(D) PORTION OF SURFACE BUILT UP BY WELDING

Desired weld　　Symbol

Fig. 23-22 Application of surfacing weld symbol to indicate surfaces built up by welding. *Courtesy American Welding Society*

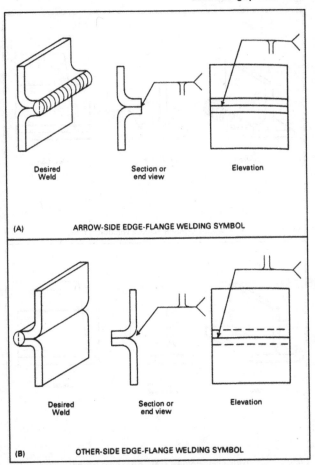

Desired Weld

Section or end view

Elevation

(A)   ARROW-SIDE EDGE-FLANGE WELDING SYMBOL

Desired Weld

Section or end view

Elevation

(B)   OTHER-SIDE EDGE-FLANGE WELDING SYMBOL

Fig. 23-23 Application of edge-flange welding symbol. *Courtesy American Welding Society*

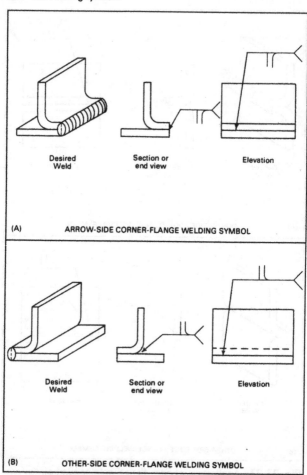

(A) ARROW-SIDE CORNER-FLANGE WELDING SYMBOL

(B) OTHER-SIDE CORNER-FLANGE WELDING SYMBOL

Fig. 23-24 Application of corner-flange welding symbol. *Courtesy American Welding Society*

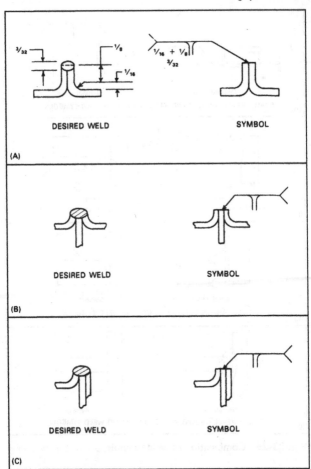

Fig. 23-25 Application of edge- and corner-flange welding symbols. *Courtesy American Welding Society*

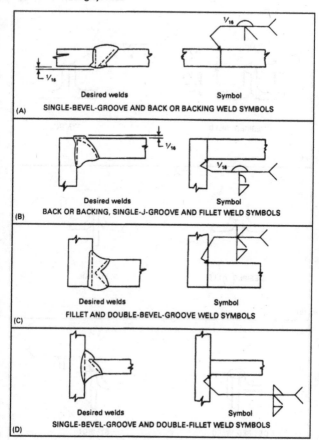

(A) SINGLE-BEVEL-GROOVE AND BACK OR BACKING WELD SYMBOLS

Desired welds   Symbol

(B) BACK OR BACKING, SINGLE-J-GROOVE AND FILLET WELD SYMBOLS

Desired welds   Symbol

(C) FILLET AND DOUBLE-BEVEL-GROOVE WELD SYMBOLS

Desired welds   Symbol

(D) SINGLE-BEVEL-GROOVE AND DOUBLE-FILLET WELD SYMBOLS

Desired welds   Symbol

Fig. 23-26 Combination of weld symbols. *Courtesy American Welding Society*

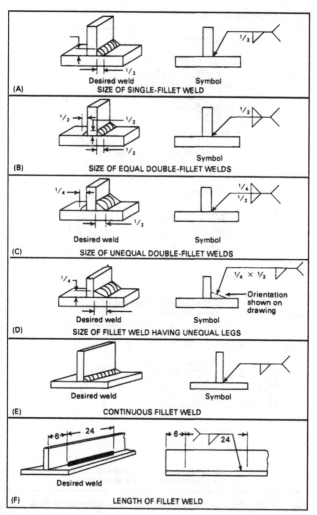

Fig. 23-27 Application of dimensions to fillet welding symbols.
Courtesy American Welding Society

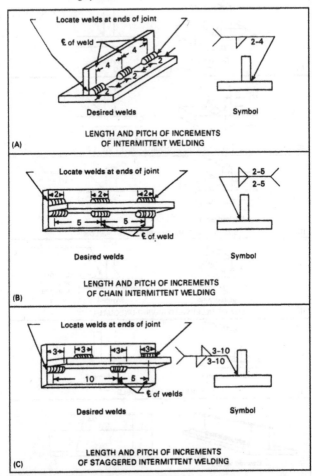

Fig. 23-28 Application of dimensions to intermittent fillet welding symbols. *Courtesy American Welding Society*

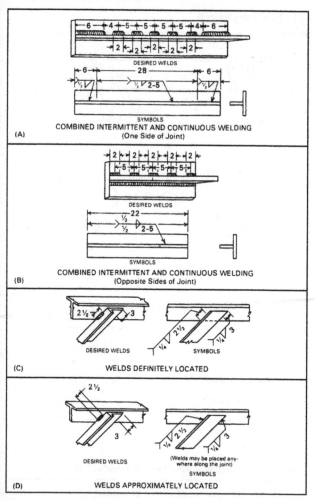

Fig. 23-29 Designation of location and extent of fillet welds.

Courtesy American Welding Society

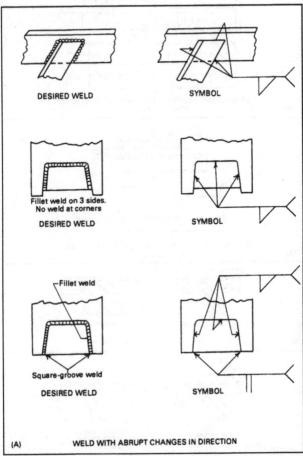

DESIRED WELD

SYMBOL

Fillet weld on 3 sides.
No weld at corners
DESIRED WELD

SYMBOL

—Fillet weld

Square-groove weld
DESIRED WELD

SYMBOL

(A)   WELD WITH ABRUPT CHANGES IN DIRECTION

Fig. 23-30 Designation of extent of welding. *Courtesy American Welding Society*

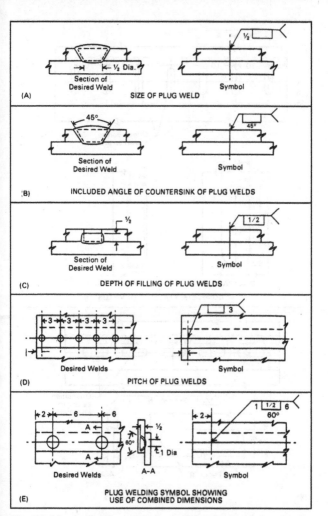

Fig. 23-31 Application of dimensions to plug welding symbols.
Courtesy American Welding Society

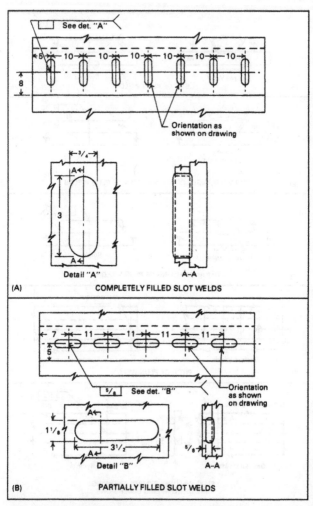

Fig. 23-32 Application of dimensions to slot welding symbols.
Courtesy American Welding Society

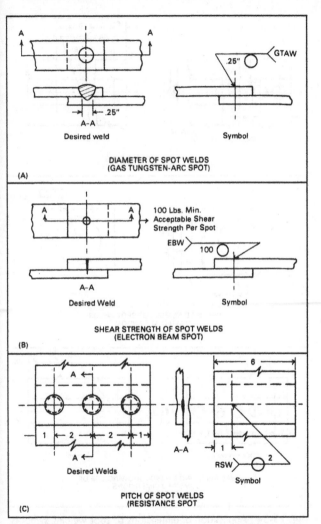

Fig. 23-33 Application of dimensions to spot welding symbols.

Courtesy American Welding Society

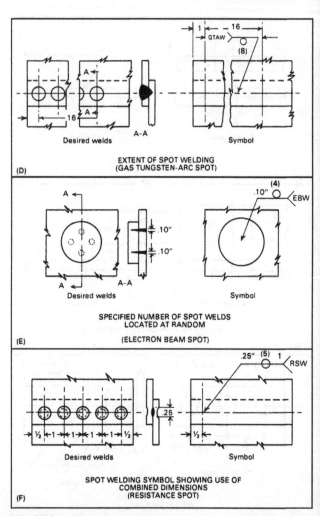

Fig. 23-34 Application of dimensions to spot welding symbols (continued). *Courtesy American Welding Society*

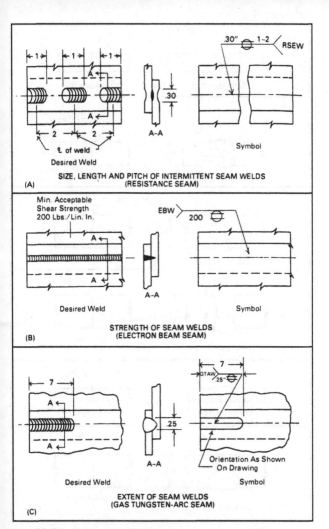

Fig. 23-35 Application of dimensions to seam welding symbols. *Courtesy American Welding Society*

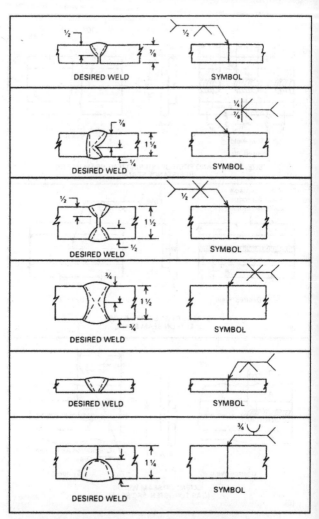

Fig. 23-36 Designation of size of groove welds with no specified root penetration. *Courtesy American Welding Society*

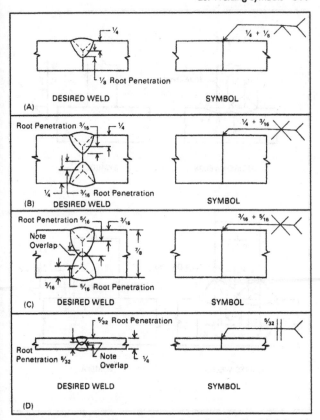

Fig. 23-37 Designation of size of groove welds with specified root penetration. *Courtesy American Welding Society*

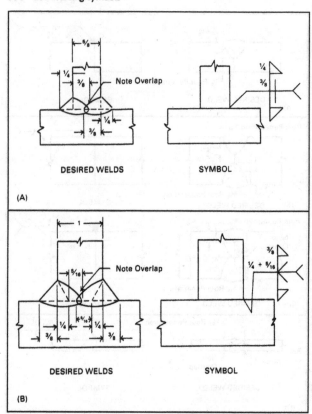

Fig. 23-38 Designation of size of combined welds with specified root penetration. *Courtesy American Welding Society*

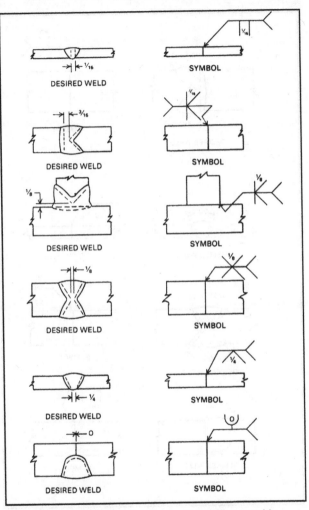

Fig. 23-39 Designation of root opening of groove welds.

Courtesy American Welding Society

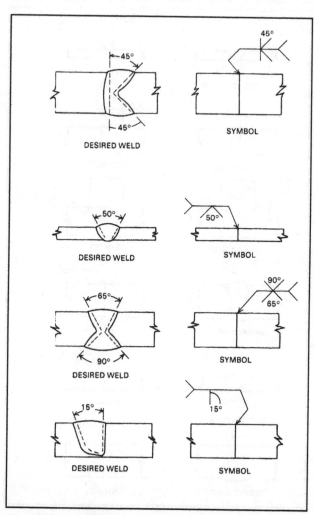

Fig. 23-40 Designation of groove angle of groove welds.

Courtesy American Welding Society

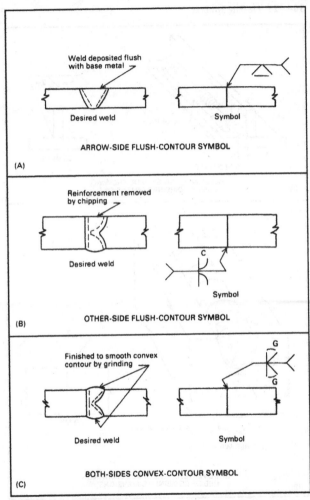

Fig. 23-41 Application of flush- and convex-contour symbols to groove welding symbols. *Courtesy American Welding Society*

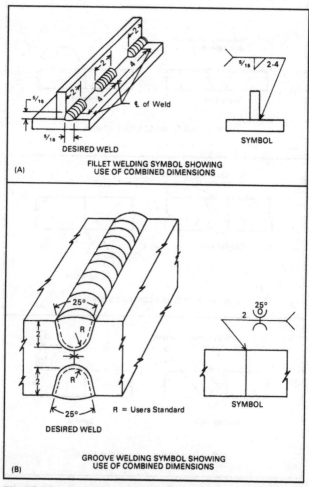

FILLET WELDING SYMBOL SHOWING
USE OF COMBINED DIMENSIONS

(A)

GROOVE WELDING SYMBOL SHOWING
USE OF COMBINED DIMENSIONS

(B)

Fig. 23-42 Application of dimensions to fillet and groove welding symbols. *Courtesy American Welding Society*

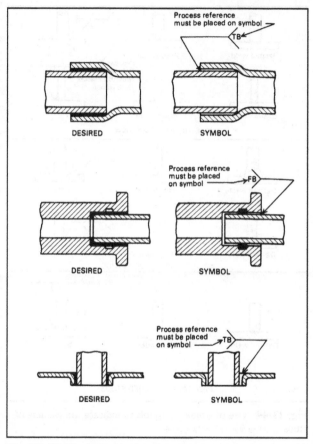

Fig. 23-43 Application of brazing symbols. *Courtesy American Welding Society*

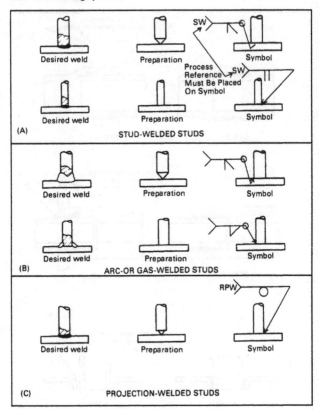

Fig. 23-44 Use of welding symbols to indicate the welding of studs. *Courtesy American Welding Society*

24. TIPS FOR PRODUCING GOOD WELDS

A good weld will be stronger than the base (parent) metal surrounding it. If an overload force is applied to the joint, the base metal will likely give before the weld does. Even the newly trained welder can consistently produce welds of high quality by exercising care in both the preparation of the weld and the welding procedure.

GENERAL RECOMMENDATIONS

- Correctly identify the base (parent) metal of the joint to be welded.

- In addition to the type of base metal, consider the design and position of the joint. Select a welding process capable of meeting these requirements.

- Use welding equipment and supplies of the highest quality. To save expense at this point is to run the risk of producing poor-quality welds.

- Choose the most suitable electrode of filler metal for the weld. This will depend on the identification of the base metal and the requirements of the welding job.

- Check and meet (or slightly exceed) the size requirements for the weld. If the metal is too thick for adequate penetration, bevel the edges to be joined.

- Whenever possible, thoroughly clean the surface of all contaminants (dirt, grease, oil, etc.). If these remain on the surface, they will likely be included in the weld and weaken it.

- Begin welding by laying the initial bead. Clean the bead, remove any weld slag, and lay the next bead (or pass) until the weld is completed.

- Examine the weld. If it is defective, remove it immediately and begin again. Never attempt to weld over a porous weld or a cracked weld.

- The finished weld should be clean and have a good appearance. All undercutting and overlapping should be repaired.

- Some metals require preheating to reduce the possibility of cracking and the formation of residual stresses. This requires correct identification of the metal being welded. If preheating is required, use only the specified preheat. Note: You should never weld on metal surfaces that are below 32°F without first preheating.

APPENDIX A
SMAW CARBON STEEL ELECTRODES

The mechanical properties and chemical analyses listed for these electrodes are nominal ones. They do vary from one manufacturer to the next for a particular electrode. Whenever possible, always check the manufacturer's electrode specifications.

Table A-1   Heavily Coated Mild Steel Electrodes

Electrode Type	Comments
E6010	• <u>General</u>: an all-purpose, fast-freeze electrode used to weld carbon and galvanized steel. It can be used for welding through light to medium amounts of dirty, rusty, or painted materials on surfaces that cannot be completely cleaned. The E6010 is recommended for welding API grades A25, A, B, and X42 pipe; for welding out-of-position X-ray welds; for general-purpose and maintenance welding; for welding pressure vessel fittings; for construction and shipbuilding welding; and for storage tank fabrication.
	• Electrode coating: Organic coating; high-cellulose sodium, 0–10% iron powder.
	• Welding current/polarity: Direct current, electrode positive (DCEP)
	• Arc type and penetration: Quick-starting, deep-penetrating, steady spray-type arc with average deposition rate.

(continued)

Table A-1 (continued)

Electrode Type	Comments
	• Welding positions: Used in all positions. Good metal transfer and rapid setup in vertical up, vertical down, and overhead welding positions.
	• Welding characteristics: Moderate spatter and moderately easy slag removal. Weld puddle wets and spreads well. Weld bead has rippled and flat appearance.
	• Typical mechanical properties: Minimum tensile strength: 70,000–79,800 psi; yield point: 61,000 psi; minimum elongation in 2.0″: 22–29%; charpy V-notch (@ −20°F) 36 lb. ft.
	• Typical chemical analysis: carbon (0.13%), manganese (0.51%), phosphorous (0.014%), sulfur (0.006%), silicon (0.20%).
E6011	• General: An all-purpose, fast-freeze electrode used to weld carbon and galvanized steel, and for rusty and oily steel in maintenance and repair work. It has the same characteristics as the E6010 electrode but can be used with both AC and DC currents. Typical applications include general-purpose repair and fabrication work; galvanized steel work; shipbuilding; structural work; welding truck frames, storage tanks, railway cars, heavy equipment, and boilers. Commonly used to weld mild steels such as ASTM A-36, A-283, A-284, A-285, A-515, and A-516.

(continued)

Table A-1 *(continued)*

Electrode Type	Comments

- Electrode coating: Organic coating; high cellulose potassium, 0% iron powder.

- Welding current/polarity: AC current. The E6011 electrode can also be used with DCEP or DCEN current, but with decreased joint penetration. AC current eliminates the arc blow problem sometimes encountered with the E6010 electrode running on DC current.

- Arc type and penetration: Strong arc with deep penetration and fine spray transfer.

- Welding positions: All positions. Fast freezing (rapid solidification) of the weld metal allows welding in the vertical and overhead positions.

- Welding characteristics: Moderate spatter, but more spatter than the E6010 electrode. Moderately easy slag removal. Note: light slag tends to eliminate slag holes. Average deposition rate. The weld deposit is free from porosity, holes, and pits. Rippled and flat bead appearance. Fillet and bead contours are flat instead of convex.

- Typical mechanical properties: Similar to E6010 electrode.

- Typical chemical analysis: : Similar to E6010 electrode.

(continued)

Table A-1 *(continued)*

Electrode Type	Comments
E6012	• <u>General</u>: Fill-freeze-type electrode used in general repair and fabrication work; galvanized steel work; shipbuilding; structural work; welding truck frames, storage tanks, railway cars, heavy equipment, and boilers. Widely used in most steel fabrication. Recommended in applications with joint fitup. E6012 capable of filling wide joint gaps. Economical to use because of its rapid welding speed and ease of use.
	• <u>Electrode coating</u>: Rutile coating; high titania sodium, 0–10% iron powder.
	• <u>Welding current/polarity</u>: AC or DCEN. Use DCEN in all positions, except when arc blow becomes a problem.
	• <u>Arc type and penetration</u>: Medium arc with medium penetration. Above-average deposition rate.
	• <u>Welding positions</u>: All positions. Most commonly used in downhill or flat (horizontal) position.
	• <u>Welding characteristics</u>: Slight spatter. Produces light but easily removable slag at fast speeds. More slag than the E6010 electrode, but less than the E6020. Produces small, smooth, and convex beads.
	• <u>Typical mechanical properties</u>: Minimum tensile strength: 67,000–69,400 psi; yield point: 55,000–62,300 psi: minimum elongation in 2.0″: 17–21%

(continued)

Table A-1 *(continued)*

Electrode Type	Comments
	• Typical chemical analysis: carbon (0.05%), manganese (0.31%), phosphorus (0.008%), sulfur (0.016%), silicon (0.12%).
E6013	• General: Fill-freeze-type electrode. General-purpose stick electrode for use on carbon steels. Wide variety of applications for light-gauge and heavy plate. Recommended for applications with poor joint fitup. Commonly used for farm equipment, metal furniture, auto bodies, machine parts, shaft buildup, and other applications where low spatter and good bead appearance are desirable.
	• Electrode coating: Rutile coating; high titania sodium, 0–10% iron powder.
	• Welding current/polarity: AC or DCEN. Use DCEN in all positions except when arc blow becomes a problem. Works well on low-voltage AC machines.
	• Arc type and penetration: Soft arc with light to medium penetration. Above-average deposition rate.
	• Welding positions: All positions. Recommended for vertical down welding.
	• Welding characteristics: Spatter is exceedingly low. Slag is easily removed and often self-cleaning on heavy plate. E6013 electrode produces a smooth and flat-to-convex weld bead with fine ripple appearance.

(continued)

Table A-1 *(continued)*

Electrode Type	Comments

- Typical mechanical properties: Minimum tensile strength: 67,000–71,000 psi; yield point: 62,000–63,000 psi; minimum elongation in 2.0″: 17–24%; charpy* V-notch (@32°F) 58 lb.ft.
- Typical chemical analysis

E6027
- General: Iron powder electrode recommended where high-quality groove and fillet welds are required. Suitable for heavy structural welding.
- Electrode coating: Iron powder, iron oxide.
- Welding current/polarity: AC, DCEN, and DCEP.
- Arc type and penetration: Medium arc with medium penetration.
- Welding positions: Horizontal fillet welds and flat position.
- Welding characteristics: Slight spatter with easy slag removal. High deposition rate, but slower than the E7024 electrode. Weld properties of E6027 superior to those of the E7024 electrode. Produces a flat-to-concave bead.
- Typical mechanical properties: Minimum tensile strength: 62,000–72,000 psi; yield point: 48,000–60,000 psi; minimum elongation in 2.0″: 25%.
- Typical chemical analysis: carbon (0.05–0.10%), manganese (0.65–0.95%), phosphorus (0.015–0.025%), sulfur (0.010–0.025%), silicon (0.010–0.025%).

(continued)

Table A-1 *(continued)*

Electrode Type	Comments
E7014	• <u>General</u>: Recommended for welding machine bases, frames, heavy sheet metal, and for general repair and maintenance work. Often recommended for applications where poor fitup is encountered or a higher deposition rate and travel speed is required.
	• <u>Electrode coating</u>: Rutile iron powder coating.
	• <u>Welding current/polarity</u>: AC, DCEN and DCEP.
	• <u>Arc type and penetration</u>: AC, DCEN and DCEP.
	• <u>Welding positions</u>: All positions.
	• <u>Welding characteristics</u>: Easily removable slag. Produces a smooth bead surface with fine ripples.
	• <u>Typical mechanical properties</u>: Minimum tensile strength: up to 70,000 psi; yield point: 67,700 psi; minimum elongation in 2.0″: 29.4%.
	• <u>Typical chemical analysis</u>: carbon (0.12%), chromium (0.041%), copper (0.012%), manganese (0.68%), phosphorous (0.021%), vanadium (0.023%), molybdenum (0.002%), nickel (0.053%), sulfur (0.012%), silicon (0.33%), iron (balance).
E7015	• <u>General</u>: A low-hydrogen electrode. Recommended for welding high-carbon, high-sulfur, and alloy steels.
	• <u>Electrode coating</u>: Low-hydrogen sodium.
	• <u>Welding current/polarity</u>: DCEP.

(continued)

Table A-1 *(continued)*

Electrode Type	Comments

- Arc type and penetration: Mild to medium penetration. Welding with a short arc is important for obtaining high-quality weld deposits. Above-average deposition rate.
- Welding positions: All positions.
- Welding characteristics: Slight spatter. Heavy, but easily removable slag. E7015 electrode produces a flat and slightly convex weld bead.
- Typical mechanical properties: Minimum tensile strength: 70,000 psi; yield point: 60,000 psi; minimum elongation in 2.0": 22%.
- Typical chemical analysis: carbon (0.08–0.13%), manganese (0.40–0.60%), phosphorus (0.04%), sulfur (0.04%), silicon (0.25%).

E7016

- General: Low-hydrogen electrode. Recommended for welding high-carbon, low-alloy sulfur.
- Electrode coating: Low-hydrogen potassium.
- Welding current/polarity: DCEP, AC.
- Arc type and penetration: Mild-to-medium penetration. Welding with a short arc is important for obtaining high-quality weld deposits.
- Welding positions: All positions.
- Welding characteristics: Slight spatter with very easy slag removal. The E7016 electrode produces a smooth and convex weld bead. Above-average deposition rate.

(continued)

Table A-1 (continued)

Electrode Type	Comments
	• Typical mechanical properties: Minimum tensile strength: 70,000–79,000 psi; yield point: 60,000–69,500 psi; minimum elongation in 2.0″: 22–28%.
	• Typical chemical analysis: carbon (0.07%), manganese (0.09%), silicon (0.50%).
E7018	• General: A low-hydrogen, iron-powder-type, fill-freeze electrode used to weld low-, medium-, and high-carbon steels. Recommended for joints involving high strength and high carbon. Commonly used for pipe, heavy sections of plate boiler work, and low-temperature equipment.
	• Electrode coating: A low-hydrogen base flux with iron powder added.
	• Welding current/polarity: DCEP or AC. Use AC current with electrode diameters larger then 5/32″. Use DCEP with diameters 5/32″ and smaller. Note: not recommended for low-voltage AC welders.
	• Arc type and penetration: Smooth, quiet arc and medium arc penetration. Note: as is common with all low-hydrogen electrodes, a short arc length should be maintained at all times. E7018 electrodes require a drag to 1/16″ maximum arc length to obtain the desired weld quality and mechanical properties. Porosity and a deterioration of impact properties may occur with an arc length of 1/8″ or longer.

(*continued*)

Table A-1 *(continued)*

Electrode Type	Comments

- Welding positions: All positions. An E7018 electrode is often recommended for out-of-position welding and tacking. Note: fillet welds in the horizontal and flat welding positions have a slightly convex weld face.
- Welding characteristics: Slight spatter with very easy slag removal. Flat-to-convex bead appearance with a smooth and finely rippled bead surface. The E7018 electrode is very susceptible to moisture, which may lead to weld porosity. Sometimes used for the final bead or layer (for better appearance) after laying the first bead with an E6010 electrode. Medium-to-high deposition rate.
- Typical mechanical properties: Minimum tensile strength: 72,000–78,000 psi; yield point: 60,000–68,000 psi; minimum elongation in 2.0″: 22–31%; charpy V-notch (@ −20°F) 65 lb ft(see footnote).
- Typical chemical analysis: carbon (0.04%), manganese (1.06%), phosphorus (0.012%), sulfur (0.011%), silicon (0.69%).

E7024

- General: Iron powder, high-speed, heavily coated electrode. Typical applications include shipbuilding, bridge construction, structural steels, storage tanks, and truck frames.
- Electrode coating: Iron powder titania; rutile type with higher efficiency (more iron powder) than type E7014.
- Welding current/polarity: AC, DCEP, and DCEN

(continued)

Table A-1 (continued)

Electrode Type	Comments
	• <u>Arc type and penetration:</u> Mild penetration. Increased penetration with little or no root porosity in horizontal or positioned fillets.
	• <u>Welding positions:</u> Suitable for use in horizontal (flat) position and for standing fillets.
	• <u>Welding characteristics:</u> Slight spatter. Easy slag removal. Often self-cleaning. Produces smooth and convex bead appearance. Very high deposition rate.
	• <u>Typical mechanical properties:</u> Minimum tensile strength: 72,000–81,000 psi; yield point: 60,000–71,000 psi: minimum elongation in 2.0″: 17–26%.
	• <u>Typical chemical analysis:</u> carbon (0.06%), manganese (0.81%), phosphorus (0.018%), sulfur (0.019%), silicon (0.43%).
E7028	• <u>General:</u> Low-hydrogen, fast-fill-type electrode used in high-production, fast-deposition-rate welding.
	• <u>Electrode coating:</u> A low-hydrogen based flux with iron powder added. Coating contains a higher percentage of iron powder (50%) than the E7018 electrode. Thicker and heavier coating than the E7018.
	• <u>Welding current/polarity:</u> AC or DCEP.
	• <u>Arc type and penetration:</u> Shallow to mild penetration. Welding with a short arc is recommended for obtaining high-quality weld deposits.

(continued)

Table A-1 *(continued)*

Electrode Type	Comments

- Welding positions: Horizontal and flat positions. Recommended for flat fillets and deep groove joints.

- Welding characteristics: Slight spatter with very easy slag removal. Heavy slag peels off. Smooth and slightly convex bead appearance. Excellent restriking for skip and tack welding. Fastest deposition rate among all the electrodes.

- Typical mechanical properties: Minimum tensile strength: 72,000 psi; yield point: 60,000 psi; minimum elongation in 2.0": 22%.

- Typical chemical analysis: carbon (0.04%), manganese (1.06%), phosphorus (0.012%), sulfur (0.011%), silicon (0.69%).

*Charpy. Testing done with notched specimens. Such tests give results that are more accurately described as notch toughness. The two standard tests are Izod and Charpy. In recent years, charpy has been replaced by Izod.

APPENDIX B
SMAW STAINLESS STEEL ELECTRODES

Table B-1 Common SMAW Stainless Steel Electrodes

Electrode Type	Comments
E308-15, E308-16	• Sound weld metal and corrosion resistance equal to or greater than that of the base metal. • Ground and polished welds cannot be distinguished from the base metal. • Designed for welding 301, 302, 304, and 308 stainless steels. • Suitable for welding 18-8 stainless steels of lower alloy content. • Typical mechanical properties: minimum tensile strength 85,000–95,000 psi; minimum elongation in 2.0″: 40–50%. • Typical chemical analysis: carbon (0.07% maximum), chromium (19.0%), nickel (9.5%), manganese (1.6%), silicon (0.50%).
E308L-15, E308L-16	• Minimum of 0.04% carbon is deposited in the weld metal. • Formation of chromium carbides in weld metal minimized. • Developed for welding 304-ELC stainless steel. • Recommended for welding 321 and 347 steels. • Used extensively for welding chemical plant equipment.

(continued)

Table B-1 *(continued)*

Electrode Type	Comments

	• Typical mechanical properties: minimum tensile strength: 80,000–90,000 psi; minimum elongation in 2.0″: 40–50%.
	• Typical chemical analysis: carbon (0.04% maximum), chromium (19.0%), nickel (9.5%), manganese (1.0%), silicon (0.30%).
E309-15, E309-16	• Excellent corrosion-resistance properties at room temperatures.
	• High strength and creep values.
	• Weld metal exhibits oxidation resistance up to temperatures of 2000°F.
	• Designed for welding 309 alloy steel.
	• Mostly used for welding dissimilar metals, such as mild or carbon steel to stainless steel.
	• Typical mechanical properties: minimum tensile strength: 85,000–95,000 psi; minimum elongation in 2.0″: 45%.
	• Typical chemical analysis.
E309Cb-15, E309Cb-16	• Inhibits carbide precipitation in 309Cb alloy when maximum corrosion and oxidation resistance at elevated temperatures is required.
	• Recommended for welding Type 347 or 321 stainless clad steels.
	• Used in welding aircraft exhaust systems.
	• Typical mechanical properties: minimum tensile strength: 85,000–95,000 psi; minimum elongation in 2.0″: 30–40%.

(continued)

Table B-1 *(continued)*

Electrode Type	Comments
	• Typical chemical analysis: carbon (0.10% maximum), chromium (23.0%), nickel (13.0%), manganese (1.6%), silicon (0.50%).
E309 Mo-15	• Designed for applications requiring molybdenum with a standard 309 analysis.
	• Used primarily for welding Type 316 clad steels.
	• Typical mechanical properties: minimum tensile strength: 85,000–95,000 psi; minimum elongation in 2.0″: 45%.
	• Typical chemical analysis: carbon (0.10% maximum), chromium (23.0%), nickel (13.0%), molybdenum (2.2%), manganese (1.7%), silicon (0.50%).
E310-15, E310-16	• Weld deposit exhibits the same chemical analysis and oxidation resistance as the base plate metal.
	• The weld deposit is fully austenitic and calls for low heat during welding.
	• Preheating to approximately 300° to 500°F produces strong, crack-free welds in dissimilar steels.
	• Developed for the welding of Type 310 stainless steel.
	• General purpose electrode for welding almost every analysis of carbon and alloy steel.
	• Frequently used to weld stainless steel that has similar composition in its wrought or cast form.

(continued)

Table B-1 *(continued)*

Electrode Type	Comments
	• Typical mechanical properties: minimum tensile strength: 85,000–95,000 psi; minimum elongation in 2.0″: 35–45%.
	• Typical chemical analysis: carbon (.20% maximum), chromium (26.0%), nickel (21.0%), manganese (1.8%), silicon (0.40%).
E310Cb-15, E310Cb-16	• Designed for special applications requiring columbium with 25% chromium and 20% nickel.
	• Especially recommended for welding Type 347 and 321 clad steels.
	• Typical mechanical properties: minimum tensile strength: 85,000–95,000 psi; minimum elongation in 2.0″: 30–40%.
	• Typical chemical analysis: carbon (0.12% maximum), chromium (26.0%), nickel (21.0%), columbium (0.80%), manganese (1.8%), silicon (0.40%).
E310Mo-15, E310Mo-16	• Recommended for welding Type 316 clad steels.
	• Used for welding other grades of molybdenum steels.
	• Typical mechanical properties: minimum tensile strength: 85,000–95,000 psi; minimum elongation in 2.0″: 35–45%.
	• Typical chemical analysis: carbon (0.12% maximum), chromium (26.0%), nickel (21.0%), molybdenum (2.0%), manganese (1.8%), silicon (0.40%).

(continued)

Table B-1 (continued)

Electrode Type	Comments
E312-15, E312-16	• Recommended for welding Type 316 clad steels. • Used for welding other grades of molybdenum steels. • Typical mechanical properties: minimum tensile strength: 110,000–120,000 psi; elongation in 2.0″: 22–75%. • Typical chemical analysis: carbon (0.15% maximum), chromium (29.0%), nickel (9.5%), manganese (1.9%), silicon (0.50%).
E316-15, E316-16	• Contains sufficient levels of chromium and nickel to render it austenitic. • Addition of molybdenum increases corrosion resistance to pitting, which is induced by such corrosive media as sulfuric acid and sulfurous acids, sulfites, chloride and cellulose solutions. • Designed for welding Type 316 chromium-nickel steel. • Used for welding wrought and cast forms of similar alloys. • Widely used in the rayon, dye, and paper industries. • Typical mechanical properties: minimum tensile strength: 85,000–95,000 psi; minimum elongation in 2.0″: 35–45%. • Typical chemical analysis: carbon (0.07% maximum), chromium (18.0%), nickel (13.0%), molybdenum (2.25%), manganese (1.7%), silicon (0.40%).

(continued)

Table B-1 *(continued)*

Electrode Type	Comments
E316L-15, E316L-16	• Controlled carbon content (0.04% maximum) reduces the possibility of the formation of intergranular carbide precipitation. • Used to weld Type 316-L or Type 318 steels. • Widely used for welding chemical equipment. • Typical mechanical properties: minimum tensile strength: 80,000–90,000 psi; minimum elongation in 2″: 35–45%. • Typical chemical analysis: carbon (0.04% maximum), chromium (18.0%), nickel (13.0%), molybdenum (2.25%), manganese (1.0%), silicon (0.30%)
E317-15, E317-16	• High molybdenum content reduces the susceptibility to pitting. • Addition of molybdenum increases the strength of this chromium-nickel alloy at elevated temperatures. • Recommended for welding Type 317 alloy steel when maximum corrosion resistance is required. • Recommended for welding air-hardenable steels. • Typical mechanical properties: minimum tensile strength: 85,000–95,000 psi; minimum elongation in 2.0″: 35–45%. • Typical chemical analysis: carbon (0.07% maximum), chromium (19.0%), nickel (13.0%), molybdenum (3.50%), manganese (1.7%), silicon (0.50%).

(continued)

Table B-1 *(continued)*

Electrode Type	Comments
E318-15, E318-16	• Columbium content eliminates formation of chromium carbides. • Recommended for welding Type 318 stainless steel when the complete absence of chromium carbides is essential. The absence of these carbides prevents weld failure caused by intergranular corrosion. • Typical mechanical properties: minimum tensile strength: 85,000–95,000 psi; minimum elongation in 2.0″: 30–40%. • Typical chemical analysis: carbon (0.07% maximum), chromium (18.0%), nickel (12.0%), molybdenum (2.25%), manganese (1.6%), silicon (0.60%).
E330-15, E330-16	• Weld deposit is highly resistant to corrosion and oxidation. • Used primarily for the fabrication of castings designed to withstand corrosion and oxidation at extreme temperature ranges. • Typical mechanical properties: Minimum tensile strength: 75,000–85,000 psi; minimum elongation in 2.0″: 25–35%. • Typical chemical analysis: carbon (0.25% maximum), chromium (15.0%), nickel (35.0%), manganese (1.6%), silicon (0.30%).
E347-15, E347-16	• Weld deposit contains columbium in the amount of ten times the carbon content, with a maximum of 1%. • Designed for welding Type 347 and 321 alloys.

(continued)

Table B-1 *(continued)*

Electrode Type	Comments
	• Developed to prevent carbide precipitation.
	• Typical mechanical properties: minimum tensile strength: 85,000–95,000 psi; minimum elongation in 2.0″: 35–45%.
	• Typical chemical analysis: carbon (0.07% maximum), chromium (19.0%), nickel (9.5%), columbium (0.80%), manganese (1.6%), silicon (0.60%).
E349-15, E349-16	• Weld deposit has excellent physical properties, including high strength-to-rupture at elevated temperatures.
	• Weld deposit is practically free of harmful carbon in the as-welded condition.
	• Structure of weld metal is extremely fine-grained.
	• Designed for welding material of similar analysis on turbojet engines.
	• Typical mechanical properties: minimum tensile strength: 105,000–110,000 psi; yield point: 80,000–85,000 psi; minimum elongation in 2.0″: 27–37%.
	• Typical chemical analysis: carbon (0.13% maximum), chromium (19.0%), nickel (9.0%), tungsten (1.40%), columbium (1.00%), molybdenum (0.50%), manganese (1.5%), silicon (0.70%).
E410-15-15, E410-16	• Weld deposit is not subject to corrosion resistance caused by carbide precipitation.

(continued)

Table B-1 *(continued)*

Electrode Type	Comments
	• Recommended for welding Type 410 straight chromium steel.
	• Used extensively for corrosion and oxidation resistance at elevated temperatures up to 1500°F.
	• Typical mechanical properties: minimum tensile strength: 80,000–90,000 psi; yield point: 55,000–60,000 psi; minimum elongation in 2.0″: 30–35%.
	• Typical chemical analysis: carbon (0.10% maximum), chromium (12.5%), manganese (0.60%), silicon (0.40%).
E430-15, E430-16	• Weld deposit is highly resistant to chemical corrosion and oxidation, up to 1600°F.
	• Recommended for welding Type 430 straight chromium steel.
	• Used for welding Type 410 steel when the chromium content of this plate is on the high side.
	• Typical mechanical properties: minimum tensile strength: 75,000–80,000 psi; yield point: 40,000–45,000 psi; minimum elongation in 2.0″: 30–35%.
	• Typical chemical analysis: carbon (0.10% maximum), chromium (16.0%), manganese (0.60%), silicon (0.60%).

APPENDIX C
SMAW ALUMINUM ELECTRODES

The aluminum electrodes covered in this appendix are the most widely used ones for welding bare, wrought, and cast aluminum. A thorough examination of aluminum electrodes is contained in the AWS publication *Specification for Bare Aluminum and Aluminum-Alloy Welding Electrodes and Rods* (AWS A5.10/A5.10M:1999). This ANSI-approved standard was prepared by the AWS Committee on Filler Metals and Allied Materials. It provides updated standards and requirements for the classification of bare, wrought, and cast aluminum-alloy electrodes and rods for use with gas metal arc (SMAL), gas tungsten arc (GTAW), oxyfuel gas (OFW), and plasma arc welding (PAW) processes.

Table C-I   SMAW Aluminum Electrodes

Electrode Type	Comments
ER1100	• Widely used general-purpose covered electrode.
	• Commonly used for welding 1100 and 3003 aluminum sheets, plates, and shapes.
	• Very resistant to chemical attack and weathering.
	• Recommended for welding aluminum storage tanks, food containers, liquid oxygen containers, pressure vessels, heat exchangers, window frames, and food processing equipment.
	• Melting point: 1190–1215° (643–657°C).
	• Tensile strength (as welded): 13,500 psi, average.

(continued)

Table C-1 *(continued)*

Electrode Type	Comments
	• Typical chemical analysis: Al (99.0 min), Be (.0008% max), Cu (0.05–0.20%), Fe and Si (.95% max), Mn (.05% max), Zn (.10% max), others (.15% max).
ER4043	• 5% aluminum-silicon filler metal.
	• Widely used general-purpose electrode for welding cast and wrought aluminum alloys.
	• Recommended for welding 2014, 3003, 3004, 5052, 6061, 6101 and 6063.
	• Melting point: 1065–1170°F (574–632°C)
	• Tensile strength (as welded): 29,000 psi, average.
	• Typical chemical analysis: Al (balance), Be (.0008% max), Cu (.30% max), Fe (.80% max), Mg (.05% max), Mn (.05% max), Si (4.5–6.0%), Ti (.20% max), others (.15% max).
ER4047/BAlSi-4	• 12% aluminum-silicon brazing rod.
	• Recommended for torch brazing and dip or furnace brazing of 1060, 1350, 3003, 5005, 6061, 6063, and 7005.
	• Melting point: 1070–1080°F (577–582°C)
	• Tensile strength (as welded): 27,000 psi, average.
	• Typical chemical analysis: Al (balance), Be (.0008% max), Cu (.30% max), Fe (.80% max), Mg (.10% max), Mn (.15% max), Zn (.20% max), others (.15% max).

(continued)

Table C-1 (continued)

Electrode Type	Comments
ER5183	• Recommended for welding 5083, 6061, 6063, 5086, 7005 and 7039 aluminum alloys. • Produces strong, tough welds with high impact and corrosion resistance. • Used for welding aluminum in structural and marine applications. • Melting Range: 1075–1180°F (579–638°C) • Tensile strength (as welded): 51,000 psi average. • Typical chemical analysis: Al (balance), Be (.0008% max), Cr (.05–.25%), Cu (.10% max), Fe (.40% max), Mg 4.3–5.2%), Mn (.5–1.0%), Si (.40% max), Ti (.15% max), Zn (.25% max), others (.15% max).
ER5356	• 5% magnesium aluminum electrode. • Exhibits strong corrosion resistance, especially to salt water. • Recommended for welding the 5000 series aluminum alloys (5050, 5052, 5083, 5356, 5454, and 5456). Note: a characteristic of the 5XXX series alloys is their susceptibility to stress corrosion cracking when the weld pool chemistry is greater than 3% magnesium and there is exposure to prolonged temperatures in excess of 150°F. Special alloys and tempers are often required to overcome this problem. • Melting Range: 1060–1175°F (571–635°C). • Tensile strength (as welded): 38,000 psi, average.

(continued)

Table C-1 *(continued)*

Electrode Type	Comments
	• Typical chemical analysis: Al (balance), Be (.0008% max), Cr (.05–.20%), Cu (.10% max), Fe (.40% max), Mg (4.5–5.5%), Mn (.05–.20%), Si (.25% max), Ti (.06–.20%), Zn (.10% max), others (.15% max).
ER5554	• Recommended for welding the 5454 and 5456 base metals.
	• Melting range: 1115–1195°F (602–646°C).
	• Tensile strength (as welded): 46,000 psi, average.
	• Typical chemical analysis: Al (balance), Be (.0008% max), Cr (0.12% max), Mg (2.27% max), Mn (0.75% max), Ti (0.12% max).
ER5556	• Recommended for welding 5083, 5086, 5154, and 5456 high tensile aluminum alloys.
	• Exhibits excellent workability and weld-ability.
	• High corrosion resistance, toughness, and strength.
	• Melting range: 1055–1175°F (568–635°C).
	• Tensile strength (as welded): 46,000 psi, average.
	• Typical chemical analysis: Al (balance), Be (.0008% max), Cr (.05–0.2%), Cu (.10% max), Fe (.40% max), Mg (4.7–5.5%), Mn (0.5–1.0%), Si (0.25% max), Ti (0.05–0.20%), Zn (0.25% max), others (.0.15% max).

APPENDIX D
CONVERSION TABLES

Table D-1 Decimal and Millimeter Equivalents of Fractional Parts of an Inch

Inches		Inches	mm	Inches		Inches	mm
	1/64	0.01563	0.397		33/64	0.51563	13.097
1/32		0.03125	0.794	17/32		0.53125	13.097
	3/64	0.04688	1.191		35/64	0.54688	13.890
1/16		0.0625	1.587	9/16		0.5625	14.287
	5/64	0.07813	1.984		37/64	0.57813	14.684
3/32		0.09375	2.381	19/32		0.59375	15.081
	7/64	0.10938	2.778		39/64	0.60938	15.478
1/8		0.125	3.175	5/8		0.625	15.875
	9/64	0.14063	3.572		41/64	0.64063	16.272
5/32		0.15625	3.969	21/32		0.65625	16.669
	11/64	0.17188	4.366		43/64	0.67188	17.065
3/16		0.1875	4.762	11/16		0.6875	17.462
	13/64	0.20313	5.159		45/64	0.70313	17.859
7/32		0.21875	5.556	23/32		0.71875	18.256
	15/64	0.23438	5.953		47/64	0.73438	18.653
1/4		0.25	6.350	3/4		0.75	19.050
	17/64	0.26563	6.747		49/64	0.76563	19.447
9/32		0.28125	7.144	25/32		0.78125	19.844
	19/64	0.29688	7.541		51/64	0.79688	20.240
5/16		0.3125	7.937	13/16		0.8125	20.637
	21/64	0.32813	8.334		53/64	0.82813	21.034
11/32		0.34375	8.731	27/32		0.84375	21.431
	23/64	0.35938	9.128		55/64	0.85938	21.828
3/8		0.375	9.525	7/8		0.875	22.225
	25/64	0.39063	9.922		57/64	0.89063	22.622
13/32		0.40625	10.319	29/32		0.90625	23.019
	27/64	0.42188	10.716		59/64	0.92188	23.415
7/16		0.4375	11.113	15/16		0.9375	23.812
	29/64	0.45313	11.509		61/64	0.95313	24.209
15/32		0.46875	11.906	31/32		0.96875	24.606
	31/64	0.48438	12.303		63/64	0.98438	25.003
1/2		0.5	12.700	1		1.00000	25.400

Table D-2 Decimal Inch Equivalents of Millimeters and Fractional Parts of Millimeters

mm	Inches	mm	Inches	mm	Inches	mm	Inches
1/100	0.00039	33/100	0.01299	64/100	0.02520	95/100	0.03740
2/100	0.00079	34/100	0.01339	65/100	0.02559	96/100	0.03780
3/100	0.00118	35/100	0.01378	66/100	0.02598	97/100	0.03819
4/100	0.00157	36/100	0.01417	67/100	0.01638	98/100	0.03858
5/100	0.00197	37/100	0.01457	68/100	0.02677	99/100	0.03898
6/100	0.00236	38/100	0.01496	69/100	0.02717	1	0.03937
7/100	0.00276	39/100	0.01535	70/100	0.02756	2	0.07874
8/100	0.00315	40/100	0.01575	71/100	0.02795	3	0.11811
9/100	0.00354	41/100	0.01614	72/100	0.02835	4	0.15748
10/100	0.00394	42/100	0.01654	73/100	0.02874	5	0.19685
11/100	0.00433	43/100	0.01693	74/100	0.02913	6	0.23622
12/100	0.00472	44/100	0.01732	75/100	0.02953	7	0.27559
13/100	0.00512	45/100	0.01772	76/100	0.02992	8	0.31496
14/100	0.00551	46/100	0.01811	77/100	0.03032	9	0.35433
15/100	0.00591	47/100	0.01850	78/100	0.03071	10	0.39370
16/100	0.00630	48/100	0.01890	79/100	0.03110	11	0.43307
17/100	0.00669	49/100	0.01929	80/100	0.03150	12	0.47244
18/100	0.00709	50/100	0.01969	81/100	0.03189	13	0.51181
19/100	0.00748	51/100	0.02008	82/100	0.03228	14	0.55118
20/100	0.00787	52/100	0.02047	83/100	0.03268	15	0.59055
21/100	0.00827	53/100	0.02087	84/100	0.03307	16	0.62992
22/100	0.00866	54/100	0.02126	85/100	0.03346	17	0.66929
23/100	0.00906	55/100	0.02165	86/100	0.03386	18	0.70866
24/100	0.00945	56/100	0.02205	87/100	0.03425	19	0.74803
25/100	0.00984	57/100	0.02244	88/100	0.03465	20	0.78740
26/100	0.01024	58/100	0.02283	89/100	0.03504	21	0.82677
27/100	0.01063	59/100	0.02323	90/100	0.03543	22	0.86614
28/100	0.01102	60/100	0.02362	91/100	0.03583	23	0.90551
29/100	0.01142	61/100	0.02402	92/100	0.03622	24	0.94488
30/100	0.01181	62/100	0.02441	93/100	0.03661	25	0.98425
31/100	0.01220	63/100	0.02480	94/100	0.03701	26	1.02362

Table D-3 Metric and English Equivalents

Metric	English
1 meter	39.37 inches, or 3.28083 feet, or 1.09361 yards
0.3048 meter	1 foot
1 centimeter	0.3937 inch
2.54 centimeters	1 inch
1 millimeter	0.03937 inch, or nearly 1/25 inch
25.4 millimeters	1 inch
1 gram	15.432 grains
0.0648 gram	1 grain
28.35 grams	1 ounce avoirdupois
1 liter (1 cubic decimeter)	61.023 cubic inches
	.03531 cubic foot
	0.2642 gal. (American) 2.202 lbs. of water at 62°F.
28.317 liters	1 cubic foot
3.785 liters	1 gallon (American)
4.543 liters	1 gallon (Imperial)

Table D-4 English to Metric Conversion Table

	English	Multiplication Factor	Metric
Pressure	psi (lb/in²)	0.06895	bars
	psi (lb/in²)	6.895	kPa
	psi (lb/in²)	0.006895	MPa
	psi (lb/in²)	0.0703	kg/cm²
Flow Rate	ft³/hr	0.472	liters/min
Length	in.	25.40	mm
Weight	lb	0.453592	kg
Capacity	ft³	0.028316	m³
	in³	16.39	cm³
	in³	0.01639	liters
	gallons (US)	3.785	liters

Table D-5 Metric to English Conversion Table

	Metric	Multiplication Factor	English
Pressure	bars	14.5	psi (lb/in^2)
	kPa	0.145	psi (lb/in^2)
	MPa	145	psi (lb/in^2)
	kg/cm^2	14.224	psi (lb/in^2)
	kg/cm^2	32.843	Ft of water (60°F)
Flow Rate	liters/min	2.12	ft^3/hr
Length	mm	0.03937	in.
Weight	kg	2.20462	lb
Capacity	m^3	35.315	ft^3
	cm^3	0.06102	in^3
	liters	61.02	in^3
	liters	0.2642	gallons (US)

Table D-6 Equivalent Temperature Readings for Fahrenheit and Celsius Scales

°F	°C	°F	°C	°F	°C	°F	°C
−459.4	−273.0	−21.0	−29.4	17.6	−8.0	56.0	13.3
−436.0	−270.0	−20.0	−29.0	18.0	−7.8	57.0	13.9
−418.0	−260.0	−20.0	−28.9	19.0	−7.2	57.2	14.0
−400.0	−240.0	−19.0	−28.3	19.4	−7.0	58.0	14.4
−382.0	−230.0	−18.4	−28.0	20.0	−6.7	59.0	15.0
−364.0	−220.0	−18.0	−27.8	21.0	−6.1	60.0	15.6
−346.0	−210.0	−17.0	−27.2	21.2	−6.0	60.8	16.0
−328.0	−200.0	−16.6	−27.0	22.0	−5.6	61.0	16.1
−310.0	−190.0	−16.0	−26.7	23.0	−5.0	62.0	16.7
−292.0	−180.0	−15.0	−26.1	24.0	−4.4	62.6	17.0
−274.0	−170.0	−14.8	−26.0	24.8	−4.0	63.0	17.2
−256.0	−160.0	−14.0	−25.6	25.0	−3.9	64.0	17.8
−238.0	−150.0	−13.0	−25.0	26.0	−3.3	64.4	18.0

(continued)

Table D-6 (continued)

°F	°C	°F	°C	°F	°C	°F	°C
−220.0	−140.0	−12.0	−24.4	26.6	−3.0	65.0	18.3
−202.0	−130.0	−11.2	−24.0	27.0	−2.8	66.0	18.9
−184.0	−120.0	−11.0	−23.9	28.0	−2.2	66.2	19.0
−166.0	−110.0	−10.0	−23.3	28.4	−2.0	67.0	19.4
−148.0	−100.0	−9.4	−23.0	29.0	−1.7	68.0	20.0
−139.0	−95.0	−9.0	−22.8	30.0	−1.1	69.0	20.6
−130.0	−90.0	−8.0	−22.2	30.2	−1.0	69.8	21.0
−121.0	−85.0	−7.6	−22.0	31.0	−0.6	70.0	21.1
−112.0	−80.0	−7.0	−21.7	32.0	0.0	71.0	21.7
−103.0	−75.0	−6.0	−21.1	33.0	+0.6	71.6	22.0
−94.0	−70.0	−5.8	−21.0	33.8	1.0	72.0	22.2
−85.0	−65.0	−5.0	−20.6	34.0	1.1	73.0	22.8
−76.0	−60.0	−4.0	−20.0	35.0	1.7	73.4	23.0
−67.0	−55.0	−3.0	−19.4	35.6	2.0	74.0	23.3
−58.0	−50.0	−2.2	−19.0	36.0	2.2	75.0	23.9
−49.0	−45.0	−2.0	−18.9	37.0	2.8	75.2	24.0
−40.0	−40.0	−1.0	−18.3	37.4	3.0	76.0	24.4
−39.0	−39.4	−0.4	−18.0	38.0	3.3	77.0	25.0
−38.2	−39.0	0.0	−17.8	39.0	3.9	78.0	25.6
−38.0	−38.9	+1.0	−17.2	39.2	4.0	78.8	26.0
−37.0	−38.3	1.4	−17.0	40.0	4.4	79.0	26.1
−36.4	−38.0	2.0	−16.7	41.0	5.0	80.0	26.7
−36.0	−37.8	3.0	−16.1	42.0	5.6	80.6	27.0
−35.0	−37.2	3.2	−16.0	42.8	6.0	81.0	27.2
−34.6	−37.0	4.0	−15.6	43.0	6.1	82.0	27.8
−34.0	−36.7	5.0	−15.0	44.0	6.7	82.4	28.0
−33.0	−36.1	6.0	−14.4	44.6	7.0	83.0	28.3
−32.8	−36.0	6.8	−14.0	45.0	7.2	84.0	28.9
−32.0	−35.6	7.0	−13.9	46.0	7.8	84.2	29.0
−31.0	−35.0	8.0	−13.3	46.4	8.0	85.0	29.4
−30.0	−34.4	8.6	−13.0	47.0	8.3	86.0	30.0
−29.2	−34.0	9.0	−12.8	48.0	8.9	87.0	30.6
−29.0	−33.9	10.0	−12.2	48.2	9.0	87.8	31.0

(continued)

Table D-6 *(continued)*

°F	°C	°F	°C	°F	°C	°F	°C
−28.0	−33.3	10.4	−12.0	49.0	9.4	88.0	31.1
−27.4	−33.0	11.0	−11.7	50.0	10.0	89.0	31.7
−27.0	−32.8	12.0	−11.1	51.0	10.6	89.6	32.0
−26.0	−32.2	12.2	−11.0	51.8	11.0	90.0	32.2
−25.6	−32.0	13.0	−10.6	52.0	11.1	91.0	32.8
−25.0	−31.7	14.0	−10.0	53.0	11.7	91.4	33.0
−24.0	−31.1	15.0	−9.4	53.6	12.0	92.0	33.3
−23.8	−31.0	15.8	−9.0	54.0	12.2	93.0	33.9
−23.0	−30.6	16.0	−8.9	55.0	12.8	93.2	34.0
−22.0	−30.0	17.0	−8.3	55.4	13.0	94.0	34.4
95.0	35.0	134.0	56.7	172.4	78.0	211.0	99.4
96.0	35.6	134.6	57.0	173.0	78.3	212.0	100.0
96.8	36.0	135.0	57.2	174.0	78.9	213.0	100.6
97.0	36.1	136.0	57.8	174.2	79.0	213.8	101.0
98.0	36.7	136.4	58.0	175.0	79.4	214.0	101.1
98.6	37.0	137.0	58.3	176.0	80.0	215.0	101.7
99.0	37.2	138.0	58.9	177.0	80.6	215.6	102.0
100.0	37.8	138.2	59.0	177.8	81.0	216.0	102.2
100.4	38.0	139.0	59.4	178.0	81.1	217.0	102.8
101.0	38.3	140.0	60.0	179.0	81.7	217.4	103.0
102.0	38.9	141.0	60.6	179.6	82.0	218.0	103.3
102.2	39.0	141.8	61.0	180.0	82.2	219.0	103.9
103.0	39.4	142.0	61.1	181.0	82.8	219.2	104.0
104.0	40.0	143.0	61.7	181.4	83.0	220.0	104.4
105.0	40.6	143.6	62.0	182.0	83.3	221.0	105.0
105.8	41.0	144.0	62.2	183.0	83.9	222.0	105.6
106.0	41.1	145.0	62.8	183.2	84.0	222.8	106.0
107.0	41.7	145.4	63.0	184.0	84.4	223.0	106.1
107.6	42.0	146.0	63.3	185.0	85.0	224.0	106.7
108.0	42.2	147.0	63.9	186.0	85.6	224.6	107.0
109.0	42.8	147.2	64.0	186.8	86.0	225.0	107.2
109.4	43.0	148.0	64.4	187.0	86.1	226.0	107.8
110.0	43.3	149.0	65.0	188.0	86.7	226.4	108.0

(continued)

Table D-6 *(continued)*

°F	°C	°F	°C	°F	°C	°F	°C
111.0	43.9	150.0	65.6	188.6	87.0	227.0	108.3
111.2	44.0	150.8	66.0	189.0	87.2	228.0	108.9
112.0	44.4	151.0	66.1	190.0	87.8	228.2	109.0
113.0	45.0	152.0	66.7	190.4	88.0	229.0	109.4
114.0	45.6	152.6	67.0	191.0	88.3	230.0	110.0
114.8	46.0	153.0	67.2	192.0	88.9	231.0	110.6
115.0	46.1	154.0	67.8	192.2	89.0	231.8	111.0
116.0	46.7	154.4	68.0	193.0	89.4	232.0	111.1
116.6	47.0	155.0	68.3	194.0	90.0	233.0	111.7
117.0	47.2	156.0	68.9	195.0	90.6	233.6	112.0
118.0	47.8	156.2	69.0	195.8	91.0	234.0	112.3
118.4	48.0	157.0	69.4	196.0	91.1	235.0	112.8
119.0	48.3	158.0	70.0	197.0	91.7	235.4	113.0
120.0	48.9	159.0	70.6	197.6	92.0	236.0	113.3
120.2	49.0	159.8	71.0	198.0	92.2	237.0	113.9
121.0	49.4	160.0	71.1	199.0	92.8	237.2	114.0
122.0	50.0	161.0	71.7	199.4	93.0	238.0	114.4
123.0	50.6	161.6	72.0	200.0	93.3	239.0	115.0
123.8	51.0	162.0	72.2	201.0	93.9	240.0	115.6
124.0	51.1	163.0	72.8	201.2	94.0	240.8	116.0
125.0	51.7	163.4	73.0	202.0	94.4	241.0	116.1
125.6	52.0	164.0	73.3	203.0	95.0	242.0	116.7
126.0	52.2	165.0	73.9	204.0	95.6	242.6	117.0
127.0	52.8	165.2	74.0	204.8	96.0	243.0	117.2
127.4	53.0	166.0	74.4	205.0	96.1	244.0	117.8
128.0	53.3	167.0	75.0	206.0	96.7	244.4	118.0
129.0	53.9	168.0	75.6	206.6	97.0	245.0	118.3
129.2	54.0	168.8	76.0	207.0	97.2	246.0	118.9
130.0	54.4	169.0	76.1	208.0	97.8	246.2	119.0
131.0	55.0	170.0	76.7	208.4	98.0	247.0	119.4
132.0	55.6	170.6	77.0	209.0	98.3	248.0	120.0
132.8	56.0	171.0	77.2	210.0	98.9	249.0	120.6
133.0	56.1	172.0	77.8	210.2	99.0	249.8	121.0

Table D-7 Decimal Parts of an Inch

Wire Gauge No.	American or Brown & Sharpe	Birmingham or Stubs Wire	Washburn & Moen on Steel Wire Gauge	American S. & W. Co.'s Music Wire	Imperial Wire Gauge	Stubs Steel Wire	U.S. Standard For Plate
0000000	0.651354	....	0.4000	....	0.500	....	0.500
000000	0.580049	....	0.4615	0.004	0.464	....	0.46875
00000	0.516549	0.500	0.4305	0.005	4.432	....	0.43775
0000	0.460	0.454	0.3938	0.006	0.400	....	0.40625
000	0.40964	0.425	0.3625	0.007	0.372	....	0.375
00	0.3648	0.380	0.3310	0.008	0.348	....	0.31375
0	0.32486	0.340	0.3065	0.009	0.324	....	0.3125
1	0.2893	0.300	0.2830	0.010	0.300	0.227	0.28125
2	0.25763	0.284	0.2625	0.011	0.276	0.219	0.265625
3	0.22942	0.259	0.2437	0.012	0.252	0.212	0.250
4	0.20431	0.238	0.2253	0.013	0.232	0.207	0.234375
5	0.18194	0.220	0.2070	0.014	0.212	0.204	0.21875
6	0.162202	0.203	0.1920	0.016	0.192	0.201	0.203125
7	0.14428	0.180	0.1770	0.018	0.176	0.199	0.1875
8	0.12849	0.165	0.1620	0.020	0.160	0.197	0.171875
9	0.11443	0.148	0.1483	0.022	0.144	0.194	0.15625

10	0.10189	0.134	0.1350	0.024	0.128	0.191	0.140625
11	0.090742	0.120	0.1205	0.026	0.116	0.188	0.125
12	0.080808	0.109	0.1055	0.029	0.104	0.185	0.109375
13	0.071961	0.095	0.0915	0.031	0.092	0.182	0.09375
14	0.064084	0.083	0.0800	0.033	0.080	0.180	0.078125
15	0.057068	0.072	0.0720	0.035	0.072	0.178	0.0703125
16	0.050082	0.065	0.0625	0.037	0.064	0.175	0.0625
17	0.045257	0.058	0.0540	0.039	0.056	0.172	0.05625
18	0.040303	0.049	0.0475	0.041	0.048	0.168	0.050
19	0.03589	0.042	0.0410	0.043	0.040	0.164	0.04375
20	0.031961	0.035	0.0348	0.045	0.036	0.161	0.0375
21	0.028462	0.032	0.0317	0.047	0.032	0.157	0.034375
22	0.025347	0.028	0.0286	0.049	0.028	0.155	0.03125
23	0.022571	0.025	0.0258	0.051	0.024	0.153	0.028125
24	0.0201	0.022	0.0230	0.055	0.022	0.151	0.025
25	0.0179	0.020	0.0204	0.059	0.020	0.148	0.021875
26	0.01594	0.018	0.0181	0.063	0.018	0.146	0.01875
27	0.014195	0.016	0.0173	0.067	0.0164	0.143	0.0171875
28	0.012641	0.014	0.0162	0.071	0.0149	0.139	0.015625

(continued)

Table D-7 (continued)

Wire Gauge No.	American or Brown & Sharpe	Birmingham or Stubs Wire	Washburn & Moen on Steel Wire Gauge	American S. & W. Co.'s Music Wire	Imperial Wire Gauge	Stubs Steel Wire	U.S. Standard For Plate
29	0.011257	0.013	0.0150	0.075	0.0136	0.134	0.0140625
30	0.010025	0.012	0.0140	0.080	0.0124	0.127	0.0125
31	0.008928	0.010	0.0132	0.085	0.0116	0.120	0.0109375
32	0.00795	0.009	0.0128	0.090	0.0108	0.115	0.01015625
33	0.00708	0.008	0.0118	0.095	0.0100	0.112	0.009375
34	0.006304	0.007	0.0104	....	0.0092	0.110	0.00859375
35	0.005614	0.005	0.0095	....	0.0084	0.108	0.0078125
36	0.005	0.004	0.0090	....	0.0076	0.106	0.00703125
37	0.004453	....	0.0085	....	0.0068	0.103	0.006640625
38	0.003965	....	0.0080	....	0.0060	0.101	0.00625
39	0.003531	....	0.0075	....	0.0052	0.099	
40	0.003144	....	0.0070	....	0.0048	0.097	

424